WILD URBAN PLANTS
OF THE NORTHEAST

2010

Peter Del Tredici

Wild Urban Plants
OF THE *Northeast*

A FIELD GUIDE

Foreword by Steward T. A. Pickett

Comstock Publishing Associates   *a division of*
Cornell University Press   ITHACA *&* LONDON

First published 2010 by Cornell University Press
First printing, Cornell Paperbacks, 2010
Printed in China

LIBRARY OF CONGRESS CATALOGING-IN-PUBLICATION DATA
Del Tredici, Peter, 1945–
Wild urban plants of the Northeast : a field guide / Peter Del Tredici ; foreword by Steward T.A. Pickett.
    p. cm.
Includes bibliographical references and index.
ISBN 978-0-8014-7458-3 (pbk. : alk. paper)
1. Urban plants—Northeastern States—Identification.
2. Weeds—Northeastern States—Identification. I. Title.
QK118.D45 2010
581.75′60974—dc22

                    2009034648

Cornell University Press strives to use environmentally responsible suppliers and materials to the fullest extent possible in the publishing of its books. Such materials include vegetable-based, low-VOC inks and acid-free papers that are recycled, totally chlorine-free, or partly composed of nonwood fibers. For further information, visit our website at www.cornellpress.cornell.edu.

Paperback printing   10 9 8 7 6 5 4 3 2 1

To my grandchildren
for whom the ordinary
is extraordinary

CONTENTS

FOREWORD

Steward T. A. Pickett

You hold in your hands an important book. It makes up for a long-standing neglect of the wild plants that inhabit America's cities and suburbs, filling in the untended, unnoticed, and perhaps unvalued nooks, vacant lots, ditches, and edges of industrial properties that suffuse our metropolises. These plants get around on their own; struggle unaided for light, nutrients, and water; and thrive without our direct intervention. The biological and ecological stories they represent and the benefits they contribute to our local ecosystems are often untold, but are both fascinating and important. This book is a key to a natural history that is almost invisible to most people, even though it is right before our eyes.

More than 50% of the world's people now live in or near cities. In the United States and other wealthy countries, more than 80% of the people live in cities and suburbs, and urban life is becoming the norm even in countries that have not benefited from the profits and products of the industrial revolution. While cities and suburbs have become the dominant habitat of humankind around the globe, however, too often the wild plants that grow in these areas and the unmanaged ground they inhabit are ignored. People fail to realize that were it not for the volunteer wild vegetation in our cities, suburbs, and towns we would be ecologically less well off. This ignorance is unfortunate because these plants offer urban inhabitants easy access to nature and its lessons and benefits. Once they are better understood, our urban ecosystems can help us live healthier, more rewarding, better integrated, and less consumptive lives. All that is needed is a tool to help us see them and understand what kind of ecological work they contribute in places where we have so actively pushed aside nature and the natural. This book is that tool.

Urban places are complex mixes of the human, the built, the cultivated, and the wild. It is important to approach them as a complex whole whose components are seen to contribute to sustainability, health, beauty, and function. Yet, American ecological science has been focused on nonurban places, emphasizing floras beyond the municipal limits. It is perhaps natural for people to be fascinated by the unfamiliar and the exotic, but this biological wanderlust has left us with little appreciation for the untended green that is a real and important part of our complex urban ecosystems.

My nearly two decades of work on urban ecosystems have made it clear to me that cities provide previews of global change. The climate shifts predicted for the world at large are already the reality in major cities and metropolitan areas. Elevated temperatures and carbon dioxide levels, dry soils, and concentrations of contaminants are the norm in cities. The plants that inhabit the wild spots in cities are well adapted to these stressed and disrupted habitats; indeed, it is doubtful that the plants that characterize nonurban places would be able to do as well.

If it is to fulfill its potential, the urban wild flora must be better understood and better used. In other words, its functions, not just its categories—native, exotic, invasive, naturalized—must be appreciated by professionals and citizens alike. Understanding should come before judgment when urban wild plants are concerned. This book provides a refreshingly unprejudiced look at urban wild flora and ultimately invites us all to look for better ways to appreciate wild plants and to use them in our efforts to improve the ecology and the human life of the city.

PREFACE

The genesis of this book goes back to the spring of 2004 when I was on a field trip with my landscape architecture students from the Harvard Graduate School of Design. We were visiting Spectacle Island, a capped landfill in the middle of Boston Harbor, studying the multitude of plants that were growing there. At some point, as I was explaining how to identify a particular specimen growing alongside the pathway, I casually mentioned how unfortunate it was that we lacked the class time to learn about the weedy plants that are so abundant in the urban environment. The ecology course covered the native species, but no class covered the weeds that were all around us. One of the students, Leah Broder, immediately suggested we set up a website for the class that would help students learn to identify these plants on their own time. We speculated that we could do it in a year, in time for next spring's class, by using my slides with a bit of added text.

Back at the Design School I quickly put together a grant proposal that was funded by the Harvard Center for Innovative Teaching Technologies. With this money and a lot of help from Kevin Lau in the Media Department I was able to hire Leah and another of my students, Ken Francis, to create the EVUE (*Emergent Vegetation of the Urban Environment*) website, which, remarkably, was operational by August 2005. With a continuation of the grant for a second year and additional help from two more of my students, Sharon Komarow and Addie Pierce McManamon, I was able to expand the fledgling website to about a hundred species and add new digital images. At this point, in September 2006, the money ran out, leaving the website frozen in time. It was then that I began thinking of putting all the information and photos together into book form and doubling the number of species covered. This second phase of the project took an additional four years to complete, and this book is the final result. While I have taken precautions to ensure that the information in this book is accurate, there will inevitably be some mistakes, for which I take sole responsibility.

ACKNOWLEDGMENTS

In addition to the students mentioned above I would like to acknowledge the encouragement and assistance provided by my professional colleagues, including Randy Prostak of the University of Massachusetts, who reviewed the manuscript and photos for accuracy; Les Mehrhoff of the University of Connecticut, who helped me with the identification of several problematic species; and Alan Berger, Richard Forman, Robert France, Niall Kirkwood, Christian Werthmann, and Paula Meijerink of the Harvard Graduate School of Design, who provided important comments and ideas during the early stages of writing the book. I also want to thank Roxana and Ledlie Laughlin of Cornwall, Connecticut, for generously allowing me to use their beautiful cabin in the woods to complete the manuscript during the summer of 2008; Izabela Riano, who put the finishing touches on several of the photographs, Cynthia Silvey and Melissa Guerrero who put together the data on the city of Somerville, Massachusetts (Figure 2), and my editor at Cornell University Press, Heidi Lovette. Finally, I especially want to thank my two children, Sonya and Luke, who for years uncomplainingly put up with innumerable travel delays while I stopped to take yet another "weed" picture; and my wife, Susan Klaw, whose support during the six long years of writing this book has been unwavering.

WILD URBAN PLANTS
OF THE NORTHEAST

INTRODUCTION

The flora of today surely differs from that of five hundred or more
years ago, due largely to the influence of an increasingly complicated
civilization; may it not be of interest to record in detail the ruderals and
escapes of to-day as a prophesy of the flora of the not distant future?

—Edgar Anderson and Robert Woodson
The Species of Tradescantia, 1935

The basic goal of *Wild Urban Plants* is to help the general reader identify the plants
that grow spontaneously in the urban environment and develop an appreciation
for the role they play in making our cities more livable. The 222 plants featured in
this book are the ones that fill the vacant spaces between our roads, our homes, and
our businesses; take over neglected landscapes; and line the shores of streams, rivers, lakes, and oceans. Some of the plants are native to the region and were present
before humans drastically altered the land; some were brought intentionally or unintentionally by people; and some arrived on their own, dispersed by wind, water, or
wild animals. They grow and reproduce in the city without being planted or cared
for. They are everywhere and yet they are invisible to most people.

Given that cities are human creations and that the original vegetation that once
grew there has long since disappeared, one could argue that spontaneous plants
have become the de facto native vegetation of the city. Indeed, a basic premise of
this book is that the ecology of the city is defined not only by the cultivated plants
that require ongoing maintenance and the native species that are restricted to
protected natural areas, but also by the plants that dominate the neglected inter-
stices of the urban environment. This "wasteland" flora occupies a significant percentage of the open space in many American cities, especially those with faltering
economies. Recent research indicates that if such vegetation is left undisturbed long
enough to develop into woodlands, it can provide cities with important social and
ecological services at very little cost to taxpayers (Zipperer et al. 1997; Kowarik and
Körner 2005; Weiss et al. 2005; Mauratet 2007).

Perhaps the most well-known example of a "spontaneous" plant is *Ailanthus altissima*, or tree-of-heaven, introduced from China. Widely planted in the Northeast
in the first half of the nineteenth century, *Ailanthus* was later rejected by urban tree

Spontaneous trees, such as this tree-of-heaven (*Ailanthus altissima*), can become significant components of the urban forest.

planters as uncouth and weedy. Despite concerted efforts at eradication, the tree managed to persist by sprouting from its roots and spread by scattering its wind-dispersed seeds. Its urban niche was famously described by Betty Smith in her 1943 novel, *A Tree Grows in Brooklyn*: "There's a tree that grows in Brooklyn. Some people call it the Tree of Heaven. No matter where its seed falls, it makes a tree which struggles to reach the sky. It grows in boarded-up lots and out of neglected rubbish heaps. It grows up out of cellar gratings. It is the only tree that grows out of cement. It grows lushly . . . survives without sun, water, and seemingly without earth. It should be considered beautiful except that there are too many of it."

Although it is ubiquitous in the urban landscape, *Ailanthus* is never counted in street tree inventories because no one planted it—and consequently its contribution to making the city a more livable place goes completely unrecognized. When the mayor of New York City promised in 2007 to plant a million trees to fight global warming, he failed to realize is that if the *Ailanthus* trees already growing throughout the city were counted he would be halfway toward his goal without doing anything. And that, of course, is the larger purpose of this book: to open people's eyes to the ecological reality of our cities and appreciate it for what it is without passing judgment on it. *Ailanthus* is just as good at sequestering carbon and creating shade as our beloved native species or showy horticultural selections. Indeed, if one were to ask whether our cities would be better or worse without *Ailanthus*, the answer would clearly be the latter, given that the tree typically grows where few other plants can survive.

There is no denying the fact that many—if not most—of the plants covered in this book suffer from image problems associated with the label "weeds"—or, to use a more recent term, "invasive species." From the plant's perspective, invasiveness is just another word for successful reproduction—the ultimate goal of all organisms, including humans. From a utilitarian perspective, a weed is any plant that grows by itself in a place where people do not want it to grow. The term is a value judgment that humans apply to plants we do not like, not a biological characteristic. Calling a plant a weed gives us license to eradicate it. In a similar vein, calling a plant invasive allows us to blame it for ruining the environment when really it is humans who are actually to blame. From the biological perspective, weeds are plants that are adapted to disturbance in all its myriad forms, from bulldozers to acid rain.

Their pervasiveness in the urban environment is simply a reflection of the continual disruption that characterizes this habitat. Weeds are the symptoms of environmental degradation, not its cause, and as such they are poised to become increasingly abundant within our lifetimes.

This book was written for the ordinary urban resident who has some curiosity about the plants and animals that live in the city—the type of person who notices and admires tenacious weeds growing in the sidewalk cracks or trees growing out of building foundations. The extensive use of photographs coupled with a minimal use of botanical jargon is intended to facilitate plant identification for people without formal botanical knowledge or training. Essentially this book is a beginner's guide to urban plants that can also serve as a primer on urban vegetation for students of all levels who are studying environmental science.

What Is a Weed?

I consulted innumerable written and electronic publications about weeds and invasive species in the process of writing this book. Most focused on plants that are considered to be problems in either an agricultural context, where the issue of competition with economic crops is the primary concern, or a residential context, where an unsightly plant is growing in a place where people are trying to cultivate something else. The term *invasive species* is used to describe a plant that displaces native vegetation in natural areas in a suburban or rural context. When it comes to *spontaneous urban plants,* people's complaints are usually aesthetic (the plants are perceived as ugly signs of blight and neglect) or security related (they shield illicit human activity or provide habitat for vermin). Indeed, the context in which a plant is growing not only determines the label that we put on it but also the positive or negative value that we assign to it.

Although there is considerable overlap among the three categories of plants listed above, they can be readily distinguished by the types of landscapes in which they grow—the constructed, the agricultural, and the ecological. In general, urban habitats have an abundance of pavement, problematic soil conditions, and frequent physical disturbance; agricultural habitats combine extreme soil disturbance with high nutrient levels; and ecological habitats are characterized by relatively low levels of soil disturbance and introduced nutrients. While disturbance is clearly an integral part of the ecology of all three types of habitats, they differ from one another in the frequency with which the disturbance reorders the environment and resets the so-called successional clock.

Succession—the change in the composition of biological communities over time —is typically driven by disturbance events. The initial stages of the process are referred to as *early succession* and are dominated by rapidly growing species that do best in full sun. Over time, these plants give way to shade-tolerant *late succession* species, which persist on the site until the next round of disturbance.

Disturbance of vacant urban land tends to be periodic. Structures are built, used, abandoned, torn down, and so on. Such disturbance is often tied to governmental

Common reed dominates this "loading dock wetland" in an abandoned Detroit factory.

approval processes that typically take 5 to 20 years for implementation. For agricultural land, the typical disturbance cycle occurs on an annual basis and begins every time the ground is plowed. The disturbance cycle for woodland landscapes is on the order of 50 to 100 years, depending on a combination of unpredictable climatic, biological, and economic factors.

The duration of the "disturbance return" cycle determines the composition of the plant communities that grow in each of the three habitat types. The vegetation of urban land, with its intermediate disturbance cycle, is composed of more or less equal numbers of annuals, herbaceous perennials, and woody plants. Annual and biennial species dominate agricultural landscapes; and long-lived woody plants dominate forested landscapes (Muratet et al. 2007).

Criteria for Developing the List of Wild Urban Plants

Developing the list of plants covered in this book was a lengthy process that required me to develop a biologically relevant definition of the term *urban*, which as I use it, refers to any part of a city or town where more of the land is covered with pavement and buildings than not, and all traces of original native habitats have disappeared. The urban environment is also characterized by high levels of disturbance associated with pedestrian and vehicular traffic, infrastructure maintenance, and new construction. Plants that can survive and reproduce under such conditions are referred to as spontaneous urban vegetation. From a plant's perspective, it is the abundance of paving and disturbance rather than the density of the human population that defines the urban environment. In other words, a sidewalk crack is a sidewalk crack whether it is in a city or a suburb. For the purposes of this book, the term *urban* does not include minimally disturbed, usually protected, natural areas found within the boundaries of the city and the species list does not include

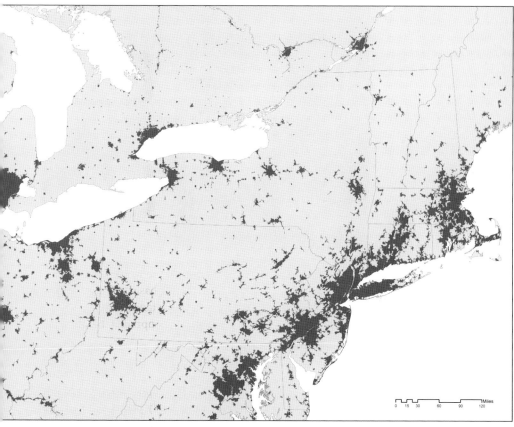

Figure 1. Urban areas of northeastern North America showing census blocks with a population density of at least 1,000 people per square mile and surrounding census blocks with an overall density of at least 500 people per square mile. U.S. data are from the 2000 census; Canada data are primarily from the mid-1990s (Government of Canada). Illustration courtesy of Alan Berger and Case Brown of the Project for Reclamation Excellence (P-REX).

plants that are confined to such remnant habitats, although it does include native species that have moved out of such natural areas and into the city. Nor does the list include cultivated species characteristic of managed landscapes within the urban environment unless they have "escaped" and become naturalized in the surrounding unmanaged areas.

The plants covered in this book are commonly found growing spontaneously in the cities of the Northeast—both large and small—from Montreal, Quebec, in the north, to Boston, Massachusetts, in the east, to Washington, D.C., in the south, and to Detroit, Michigan, in the west, with the core coverage area extending along the eastern seaboard from Boston to Philadelphia. My categorization of a plant as common is based on a review of the relevant literature on spontaneous vegetation

Quaking aspen colonizing the roof of the abandoned Central Train Depot in Detroit.

as well as my personal observations over a number of years in numerous north-eastern cities. The decision to exclude certain plants was ultimately a subjective one on my part, motivated partly by space limitations and partly by the book's focus on species that are ubiquitous in the urban environment. As one moves south and west of the core coverage area, one is more likely to find spontaneous species not covered in this book. From the ecological perspective almost all of the plants included in this book are disturbance-adapted, early successional species (Appendix 3). Excluded from the book are many familiar ground covers such as the common orange daylily, periwinkle, and lily-of-the-valley which can spread aggressively once planted in gardens but are typically not capable of moving into new areas without the help of people.

Where Do Urban Plants Come From?

Of the 163 herbaceous angiosperms (128 dicots and 35 monocots) treated in this book, 59% came to North America from Europe (including adjacent parts of Asia and North Africa), 26% are native only to North America (including Central America), 10% are native to both North America and Eurasia, and 5% originated in Eastern Asia. Remarkably, only 1 plant—bermudagrass (*Cynodon dactylon*)—appears to be native exclusively to Africa. In contrast, the 54 woody angiosperms covered here are mainly native to temperate North America (51%) and Asia (32%), with only 17% coming from Europe (Table 1). The different geographical origins of the herbaceous species and the woody species reflect the deeper evolutionary patterns that underlie the modern distribution of urban plants (Grime 2001).

The origin and global dispersal of spontaneous urban vegetation is as much a cultural as a biological phenomenon. European ecologists have traditionally had a stronger interest in the subject than their American counterparts because Europe

TABLE 1. Geographical Origin of the Plants Described in This Book

	North & Central America	Europe & Central Asia	Eastern Asia	Eurasia & North America	Africa	Totals
Ferns & Horsetails	2	—	—	2	—	4 (1.5%)
Conifers	—	—	1	—	—	1 (0.5%)
Woody Dicots	27	9	18	—	—	54 (24.5%)
Herbaceous Dicots	34	79	6	9	—	128 (58%)
Monocots	9	17	2	6	1	35 (15.5%)
Totals	72 (32.5%)	105 (47.5%)	27 (12%)	17 (7.5%)	1 (0.5%)	222

has a much longer history of urbanization. By combining sophisticated archaeological work with modern ecological research they have been able to reconstruct the complex history of their continent's urban flora. As part of this process they subdivided the category of alien plants into *archaeophytes,* which were introduced with agriculture into a given area from other parts of Europe prior to 1500, and *neophytes,* which were introduced after 1500, mainly from Asia and North and South America (Pyšek 1998; Wittig 2004). This distinction is of little relevance to American botanists, who tend to lump all European plants together as "aliens."

The dichotomy represented by these two terms reflects the long history—going back thousands of years—of moving both food and medicinal plants around *within* the Mediterranean basin (between southern Europe, North Africa, Turkey, and the Middle East); from central Asia *into* the Mediterranean basin; and from the Mediterranean region *into* central, eastern, and northern Europe. When one realizes that twenty-two of the plants covered in this book were described almost 2000 years ago by the Greek physician Dioscorides in *De Materia Medica*—a five-volume work about medicinal plants that remained in usage into the 1600s—one begins to get an inkling for just how ancient the relationship between archaeophytes and humans is (Appendix 1). The fact that we no longer use these plants to treat diseases does not diminish the significance of the role they played in the development of human culture.

The movement of plants from Europe into North America that began in the 1600s with the founding of Plymouth colony and continued through the 1800s is part of the larger story of human migration. The earliest North American settlers from Europe brought their entire lifestyle along with them—not only their own personal belongings and food for the first year, but also seeds of their crop plants, livestock and the fodder to feed them, and medicinal plants. In addition to these intentionally cultivated plants, the colonists inadvertently brought weed seeds embedded in the hay they brought for their animals and mixed in with the grains they sowed on the land they cleared. Essentially, the weeds came to North America along with the crops. In his classic book *New-England's Rarities Discovered* (1672), John Josselyn documented the presence of dozens of European weeds growing spontaneously in New England under the category "Of such plants as have sprung up since the English planted and kept cattle in New England" (Appendix 2). Clearly Europeans' invasion of North America was biological as well as cultural.

English-speaking people dominated the initial settlement of North America, but immigrants from other western European countries were quick to follow, driven by difficult economic conditions in their homelands. And just as the members of each group came with their own language and culture, they also brought their own crops and agricultural practices. Indeed, of the 150 nonnative species covered in this book, at least half were intentionally cultivated prior to their escape from the farm or garden (Rehder 1946; Mack 2000, 2003; Mack and Ernberg 2002).

This flood of European plants into North America dramatically and permanently altered the North American landscape. In contrast, only a few North American species have become naturalized in Europe, most notably black locust, black cherry, pokeweed, and box elder, along with a number of goldenrods, asters, and evening primroses; all are native to disturbed, early successional habitats (Marks 1983; Wittig 2004). This asymmetry in the biological exchange between the two continents is undoubtedly a reflection of the lob-sided nature of the cultural exchange.

After Japan was forcibly opened to the West in 1853, new plants from Asia—most notably woody plants—began pouring into North America through ports on the U.S. West Coast. At first only wealthy estate owners and commercial nurseries could afford to import these exotic species. However, with time and U.S. government support for plant exploration in China, large numbers of Asian plants were introduced into our horticultural and agricultural landscapes. Many of these species—including kudzu, multiflora rose, Japanese barberry, and various honeysuckles—were widely planted from the 1930s through the 1970s when using plants to control (a legacy of the Great Depression) was considered ecologically and economically responsible.

At the same time that exotic plants were being introduced from Europe and Asia, native American species were being shuffled around as settlers moved inland from the Atlantic Coast. Perhaps the best example of this kind of movement is the black locust, *Robinia pseudoacacia,* which originally had a distribution limited to the Appalachian and Ozark mountain ranges and now grows spontaneously throughout North America—to such an extent that many states now classify it as an invasive species. Mühlenbach's (1979) monograph on the "synantropic" flora of St. Louis documents the significant role played by railroads in moving both native and introduced plants around the country during the nineteenth and twentieth centuries, to say nothing of the fact that the railroad bed itself provided an ideal corridor for the migration of disturbance-adapted species.

At the local level, the movement of plants into cities was facilitated when soil from the countryside—usually loaded with seeds and rhizomes—was used to fill in

Spontaneous vegetation dominates this front yard where a privet hedge (*Ligustrum vulgare*) once grew.

coastal and freshwater wetlands for urban development. The most extreme example of such wholesale translocation of soil and its included seeds resulted from the imbalance of trade between Europe and North America in the nineteenth century when European ships coming to the United States typically arrived filled with rocks and soil as ballast. This ballast was then discarded on shore before loading up with cargo for the return trip—creating extensive "ballast grounds" in many American port cities. Needless to say, the novel assemblage of plants that showed up on these piles attracted the attention of observant botanists, who carefully documented the arrival and spread of the stowaways (Mehrhoff 2000).

These eighteenth- and nineteenth-century manifestations of globalization pale in comparison with what has taken place during the twentieth century as formerly independent economies have become totally interdependent. People and commercial goods now flow seamlessly around the globe, accompanied by a host of weeds, pests, and pathogens (Kareiva et al. 2007). A recent example of this kind of unintentional exchange is the Asian longhorned beetle, which entered North America in the 1990s embedded in wooden shipping pallets and now threatens maple trees growing in several metropolitan areas in the Northeast.

In much the same way that globalization has destabilized long-standing local economic institutions, it has also destabilized well-established ecological associations and led to the establishment of new ones. In essence, the plants that grow in our cities are a cosmopolitan array of species that reflect the natural and cultural histories of the area. Every type of land use seems to leave behind as a legacy a few species able to make the transition to the new type of land use. For cities, this sequence starts with native species adapted to ecological conditions before the city was built. These are followed, in succession, by species adapted to agriculture and pasturage, to pavement and compacted soil, to lawns and landscapes, to infrastructure and pollution, and ultimately to rubble and abandonment.

What Makes a Successful Urban Plant?

The plants that grow and survive in derelict urban wastelands are famous (or infamous) for their ability to grow under extremely harsh conditions. There are many reasons why some plants can survive on these sites while others die, but in essence, the plants that can survive and reproduce on their own in the urban environment are among the toughest on the planet. Through a quirk of evolutionary fate, they developed traits in their native habitats such as the formation of a taproot, that seem to have "preadapted" them to flourish in urban sidewalk cracks (Larson et al. 2004). One study concluded that many successful urban plants are native to exposed cliffs or dry, open grasslands, both of which are characterized by soils with a relatively high pH (Lundholm and Marlin 2006). Cities, with their tall, granite-faced buildings and concrete foundations, are in a sense the equivalent of the natural limestone

---

TABLE 2. **General Characteristics of Spontaneous Urban Plants**

– Their seeds can germinate under a wide variety temperature and light conditions.
– Their seeds have great longevity and can germinate after long periods of burial.
– Their seeds are widely dispersed by wind or animals.
– Their flowers are self-pollinating or pollinated by wind or generalist insects.
– They are adaptable in their growth requirements and can tolerate a wide range of nutrient, light, moisture, and temperature conditions.
– They can readily colonize disturbed soils in full sun.
– They can tolerate problematic soils characterized by any or all of the following: low levels of organic matter, high levels of chemical contamination, relatively high pH, and high levels of compaction.
– They can tolerate drought induced by the abundance of impervious pavement.
– Annual species begin flowering at an early age and produce seed over a long period of time.
– Annuals and herbaceous perennial species display a high degree of "phenotypic plasticity": when growing conditions are good they are large and produce abundant seed crops, and when conditions are poor they are stunted with limited seed production.
– Many herbaceous species display a prostrate growth habit with a strong taproot and are tolerant of trampling or mowing.
– Many perennial species have low winter chilling requirements and begin growing early in spring, as soon as weather permits. This tendency is accentuated in the city relative to the country because of the warmer winter temperatures.
– Many woody species can sprout vigorously from specialized underground structures or cut stems following traumatic damage or extreme drought.

Source: Modified from H. G. Baker, "The Evolution of Weeds," *Annual Review of Ecology and Systematics* 5 (1974): 1–24.

cliff habitats where many of these species originated. Similarly, the increased use of deicing salts on our roads and highways has resulted in the development of micro-habitats along their margins that are typically colonized by calcium-loving grass-land species adapted to limestone soils as well as salt-loving plants from coastal habitats.

Plant biologists have identified a number of characteristics that seem to pre-adapt species to succeed in highly disturbed agricultural or urban environments. In Table 2, I have adapted H. G. Baker's 1974 list of "ideal weed characteristics" to fit spontaneous urban vegetation. In general, the successful urban plant needs to be *flexible* in all aspects of its life history from seed germination through flowering and fruiting, *opportunistic* in its ability to take advantage of locally abundant resources that may be available for only a short time, and *tolerant* of the stressful growing conditions caused by an abundance of pavement and a paucity of soil.

Urban Ecology

At first glance, the term *urban ecology* might seem an oxymoron. Nevertheless, cities do have their own distinctive ecology, dominated by the needs of people and driven by socioeconomic rather than biological factors. People welcome other organisms into cities to the extent that they contribute to making the environment a more attractive, more livable, or more profitable place to be; and they vilify as weeds those organisms that flourish without their approval or assistance. Regardless of humans' preferences, an enormous variety of nonhuman life has managed to crowd into cities to form a cosmopolitan collection of organisms that is every bit as diverse as the human population itself.

The complex cultural, geological, and biological histories of most American cities have made them a patchwork of ecological habitats, each characterized by a distinctive suite of plants and animals. From a strictly functional perspective, ecologists generally classify urban land into three categories: remnant, or leftover, native landscapes; intentionally planted or managed landscapes; and ruderal, or abandoned, landscapes (Whitney 1985; Zipperer et al. 1997). The details of this classification scheme are presented in Table 3.

A spontaneous "green roof" on an abandoned Detroit factory featuring tree-of-heaven, Siberian elm, and riverbank grape.

TABLE 3. Generalized Taxonomy of Urban Landscapes in the Northeast

NATIVE (REMNANT) LANDSCAPES: Patches of natural woodlands and wetlands persisting from times of early settlement; they are dominated by **native plants** growing on relatively undisturbed soils and require low levels of maintenance.

Freshwater wetlands
River and stream corridors
Saltwater marshes
Woodlands
Cliffs and rock outcrops

MANAGED (CONSTRUCTED) LANDSCAPES: Intentionally designed planting areas that serve specific public functions; they are dominated by **cultivated plants** growing on relatively rich soil (often imported or manufactured) and require moderate to intensive maintenance to preserve their integrity.

Lawns and ball fields
Public parks and cemeteries with horticultural plantings
Residential and commercial landscapes
Developed river and stream corridors
Street trees, infrastructure plantings, and parking lots

RUDERAL (ADAPTIVE) LANDSCAPES: Abandoned or neglected land and urban infrastructure dominated by **spontaneous plants** growing on disturbed or compacted soils and require zero maintenance.

Abandoned or Degraded Open Space

Trampled lawns and ball fields
Neglected ornamental landscapes (residential, commercial, and public)
Emergent woodlands and thickets on abandoned land
Vacant lots in various states of succession
Freshwater wetlands, ponds, and streams
Channelized riverbanks (negatively impacted by roadways)
Saltwater marshes with impeded drainage
Exposed rock outcrops

Urban Infrastructure

Small pavement openings (tree pits), edges, and cracks
Chain-link fence lines
Stone and masonry walls
Alleyways (in perpetual shade)
Compacted dirt walkways
Roadway and highway banks, edges, and median strips
Railroad beds and rights-of-way (with gravel substrate)

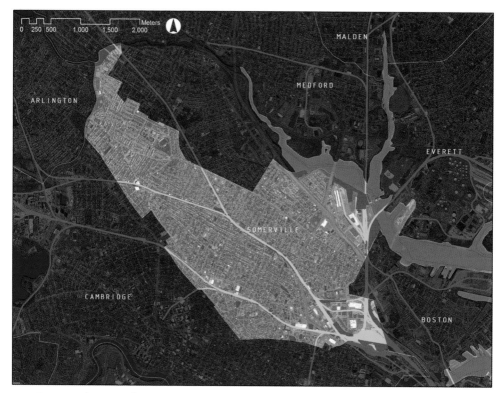

Figure 2. This map of Somerville, Massachusetts, shows vacant residential, commercial, and industrial lots (red) and railroad right-of-ways (yellow) where spontaneous plants are the dominant vegetation. The vacant parcels of land were identified on the basis of parcel-level tax assessments, and they were mapped using ESRI's ArcMap 9.2 software based on existing GIS data provided by MassGIS and the City of Somerville. The average widths of the rail corridors were calculated based on a 40-foot (12.2 m) buffer from the centerlines of existing railroads, both active and inactive. Based on the overlay of these data, the minimum area of the city of Somerville dominated by spontaneous vegetation is calculated to be roughly 9.7% of the total land area of the city (0.4 of 4.1 square miles [1.03 km² of 10.6 km²]). Map prepared by Cynthia Silvey and Melissa Guerrero with assistance from Ellen Schneider of the City of Somerville.

Essentially, remnant landscapes are left over from the time before the city spread out to embrace them. They include everything from freshwater wetlands and woodlands to salt marshes near the coast. They are residual parts of the original landscape dominated by native species growing on relatively undisturbed soils. Managed or functional urban landscapes are created specifically for human use and enjoyment. They are dominated by horticultural plants growing on relatively good soil and require a consistent input of human energy (i.e., maintenance) in order to survive. Abandoned urban landscapes receive no maintenance and are dominated by spontaneous vegetation growing mainly on compacted or fill soils. These neglected wastelands typically experience high levels of disturbance and can be as inconspicuous as a sidewalk crack or as prominent as an abandoned rail yard or vacant lot (Kastner 1993; Stalter 2004).

Disturbance Ecology

It is not a big step for a disturbance-adapted wild plant to become first an agricultural weed and then an urban weed, but neither is it a given. In the first place, the agricultural niche is much richer in terms of the availability of light, nutrients, and water than the urban environment, which is characterized by an abundance of drought-inducing pavement and compacted soil. Second, the disturbance-driven succession cycle in the city is unpredictable and depends on socioeconomic rather than seasonal factors. To borrow a term from ecology, the urban environment is *patchy* compared with the agricultural habitat. As everyone who lives in the city knows, it is always under construction: old buildings are being razed, new buildings are being erected, infrastructure is being replaced, roadways are being repaved or put underground, and, most destructive of all, open land is being cleared for commercial expansion. At any given time a significant portion of the urban fabric is in the process of being torn up and rebuilt.

Such periodic, unpredictable disturbance combined with the continual introduction of new species into the urban environment from outside sources—including nursery plants with their associated weeds, lawn-seed mixes, construction fill, and seeds carried by wind and migrating animals—provide all the components necessary to produce the typical arrested succession cycle of the urban environment. Disturbance and immigration interact on a continuous basis in the urban environment to create a constantly shifting mosaic of plant associations dominated by stress-tolerant, early successional species (Figure 3). By way of analogy, this situation is not all that different from the dynamism that characterizes the human population of the city as one ethnic group replaces another when the socioeconomic status of a given neighborhood shifts either upward or downward.

As many scientists have pointed out, modern climate change can be viewed as a massive, uncontrolled experiment on the impact of increased atmospheric carbon dioxide concentrations on the earth's ecosystem. Most people now realize that after nearly 200 years of burning fossil fuels these impacts are both wide ranging —they have affected every corner of the globe—and, at the local level, unpredictable. And this is where the cities come in.

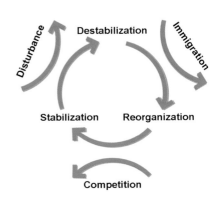

Figure 3. The drivers of the succession cycle in urban habitats.

This sea of urban blacktop looks like a volcanic lava flow, and the plants that grow here, including mullein (*Verbascum thapsus*), chicory (*Cichorium intybus*), New England hawkweed (*Hieracium saubadum*), and frost aster (*Symphyotrichum pilosum*) are extremely drought tolerant.

TABLE 4. **Common Stress Factors for Urban Plants**

- **Paving** reflects heat and light, and leads to higher temperatures in both summer and winter; its impervious nature inhibits the movement of water and air into the soil and leads to drought problems.
- **De-icing salts** used along roads and walkways elevate the soil pH and can lead to "osmotic drought."
- **Drainage problems** are often caused by the construction of buildings and roadways, which inhibit the flow of water across a site.
- **Soil compaction** is a by-product of all construction activities and heavy foot traffic; it results in poorly drained, chronically dry soils.
- **Air pollution** from the burning of fossil fuels (especially ozone and sulfur dioxide) can injure plants and reduce soil pH.
- All of these factors, acting singly or in concert, weaken plants and make them more susceptible to damage **by insects and pathogens.**

Because all the impervious paving and buildings absorb and retain heat, and all the cars, air conditioners, and electrical equipment generate heat, the annual mean temperatures of large urban areas (i.e., with populations in excess of a million people) can be up to 5.4° F (3° C) warmer than the surrounding nonurban areas; on extreme occasions the temperature differences between the city and the countryside can be as high as 21.6° F (12° C) (Sieghardt et al. 2005; George et al. 2007). This "heat island effect" means that the core areas of many of our larger cities have already warmed up to the levels predicted for the surrounding countryside 20 to 50 years from now. Urban areas are the prefect place to study how climate change will affect the environment because they have already arrived at the future.

While increased temperature is probably the most ecologically significant factor that distinguishes the city from the surrounding countryside, several other features

A robust urban forest has developed on this vacant lot in Detroit.

associated with urban areas have profound impacts (both negative and positive) on the growth of plants as well, including greater air pollution, higher concentrations of carbon dioxide, altered solar radiation regimens, altered wind patterns, altered hydrology, and decreased humidity (Gregg et al. 2003; Sukopp 2004; George et al. 2009). In addition, the soils found in urban environments—especially those that are heavily compacted or derived from construction rubble—display numerous characteristics that distinguish them from less disturbed soils (Table 4) (Byrne 2007; Godefroud et al. 2007).

Brave New Ecology

The notion that every city has a native flora that can be restored is an idea with little credibility in light of the facts that (1) most urban land has been totally transformed from what it once was; (2) the climate conditions that the original flora was adapted to no longer exist; and (3) most urban habitats are strictly human creations with no natural analogs and no indigenous flora. A native flora once grew where the city now stands, but the idea that this vegetation can somehow be restored to the site is both ecologically and evolutionarily impossible (Gould 1998). Certainly we can plant native species in the city and they will grow—but only if we provide them with the right kind of soil and maintain them the way we would any other intentionally cultivated plant. In an urban context, the concept of restoration is really just gardening dressed up to look like ecology (Janzen 1998; Del Tredici 2007). In the absence of ongoing maintenance, the default vegetation of the cities of the

Northeast is the cosmopolitan collection of plants described in this book. These are the species that establish themselves on their own and thrive without the input of human energy. In general, they are preadapted to the early successional conditions that humans create in the urban environment, and as such they can legitimately be considered its natural vegetation (Larson et al. 2004) (see Appendix 3).

In addition to the form or composition of spontaneous urban vegetation, it is also important to consider its ecological function. In minimally disturbed native habitats, the form and function of the flora and fauna are closely linked (Tallamy 2007). In the urban environment, however, form and function can be decoupled so that reasonable ecological functionality is possible with a cosmopolitan assemblage of species. A long-term study of the Baltimore ecosystem (Pickett et al. 2008) produced some unexpected conclusions in this regard:

- The urban biota is diverse.
- Urban wetlands are not nitrate sinks.
- Nitrate water pollution is higher in suburbs than in the city.
- Urban soils are not uniformly disturbed.
- Urban areas can contribute to carbon balance.
- Lawns can have beneficial social and biogeochemical functions.

The researchers conclude that "both exotic and native species have functional value in urban systems." Needless to say, conservation activists seldom—if ever—acknowledge the ecological contributions of nonnative, spontaneous urban vegetation. Indeed, the response of many regulatory agencies to these plants is to label them invasive and, in a few states, to ban their propagation and distribution. Such well-meaning attempts to control ecology are based on the assumptions that exotic species are the cause of environmental degradation and that replacing them with native species will result in a more "natural" ecological balance. While this is theoretically possible, the reality is that in the absence of intensive horticultural maintenance (i.e., planting, weeding, mowing, and watering), spontaneous vegetation will eventually come to dominate most urban landscapes. In fact, the amount of spontaneous vegetation in a given city always seems to be inversely proportional to its economic prosperity. In New York City, for example, Manhattan—with its sky-high property values—has relatively little spontaneous vegetation while Brooklyn and the Bronx are filled with it. Similarly, Detroit has become an epicenter of spontaneous vegetation as a result of the long, slow decline of the automobile industry.

Edgar Anderson noted in 1952 that many of the so-called weeds that populate North American cities were originally "dump heap" plants from Europe that gained a competitive edge by taking advantage of the nutrient-rich waste that people left behind; in ecological terms, they were exploiting an "open niche." By virtue of their long association with human culture, these "camp followers" developed traits that made them more tolerant of human-generated pollution than the native vegetation that existed on the site before it was disturbed. Casting such plants in the role of "thugs" makes it virtually impossible to recognize the positive contributions they are making to the ecology of cities. By way of analogy, people's attitudes about

Two "pollarded" American elms (*Ulmus americana*) have adapted to this chain-link fence niche in Hartford, Connecticut.

invasive species mirror the political debate about undocumented aliens. Many people complain about the problems caused by people who are in the country illegally but fail to recognize the positive contributions such people make to the economy and the society at large.

In an effort to turn this dynamic on its head, I have chosen to describe in positive terms the ecological functions of all the plants treated in this book. Clearly the longer-lived woody plants and herbaceous perennials have the capacity to perform more of these functions than short-lived annuals and biennials, which mainly serve to absorb pollutants and stabilize disturbed ground. Under the heading ECOLOGICAL FUNCTION I have created the following list of possible "ecosystem services" for each entry in this book:

- temperature reduction
- food and/or habitat for wildlife
- erosion control on slopes and disturbed ground
- stream and river bank stabilization
- nutrient absorption (mainly nitrogen and phosphorus) in wetlands
- soil building on degraded land
- tolerance of pollution or contaminated soil
- disturbance-adapted colonizer of bare ground

I do not include carbon storage and the production of oxygen in my list of ecosystem services because all plants do those things regardless of where they come from or where they are growing. It is important to keep in mind, however, that because they grow on marginal sites and require no maintenance, urban plants are probably providing a greater return in terms of carbon sequestration than many intentionally cultivated species.

I also do not treat biodiversity per se as an ecological function. An exhaustive literature review of the vegetation of 54 Central European cities determined that they contained between 20% and 60% nonnative species, with a mean value of 40.3%. This figure is 13.7% higher than the ratio of nonnative to native plants in the surrounding region, which is indicative of a "remarkable concentration of aliens in urban areas" (Pyšek 1998). Other vegetation surveys of European cities indicate that the ratio of nonnative to native species as well as the ratio of neophytes to archaeophytes increase as one gets closer to the more highly disturbed parts of the city (Kovarik 1990; Pyšek et al. 2004).

The unexpectedly high species richness of the inner city is the product of a number of factors, including habitat heterogeneity, changing climatic conditions, horticultural activity by humans, and the establishment opportunities disturbance provides for new immigrants (Zerbe et al. 2003; Sukopp 2004). A number of European researchers have even proposed that certain types of inner-city landscapes should be conserved because of the role they play in promoting and maintaining urban biodiversity (Maurer et al. 2000; Kowarik and Körner 2005; Muratet et al. 2007).

One potentially important function of spontaneous urban vegetation that people have only recently begun to study is *phytoremediation* or the ability of some plants to clean up contaminated sites by selectively absorbing and storing high concentrations of heavy metals such as cadmium, lead, copper, zinc, chromium, and nickel in their tissues (Porębska and Ostrowska 1999). Several common urban species—most notably prickly lettuce (*Lactuca serriola*), lambsquarters *Chenopodium album*), and mugwort (*Artemisia vulgaris*)—possess this ability and are helping to "detoxify" the land by taking some of the heavy metals out of circulation.

As every sufferer of hay fever knows all too well, plants do not always enhance the quality of life for the human inhabitants of cities. Indeed, if recent research is any guide, climate change could well make some of these negative interactions between plants and people worse than they currently are. Controlled experiments with two infamous native plants—ragweed (*Ambrosia artemisiifolia*) and poison ivy (*Toxicodendron radicans*)—have shown that elevated levels of carbon dioxide induce the former to produce significantly more of its highly allergenic pollen and cause the latter to produce higher concentrations of its rash-producing toxins (Ziska 2003; Mohen et al. 2006). While the remarkable ability of these two species to adapt to changing environmental conditions does not bode well for human health in a carbon dioxide–rich future, such examples do provide an important reminder of the innate capacity of "weeds" to capitalize on the mess we have made of the planet.

It is a foregone conclusion that the environment will continue to deteriorate over the next few decades as people continue to pump more heat-trapping carbon dioxide into the atmosphere and more acid rain falls back to earth to pollute the water and the soil. The worldwide migration of people from the countryside into cities is also contributing to environmental degradation because land that was once covered with vegetation is being covered instead by buildings and pavement that generate and retain heat (Grimm et al. 2008). The confluence of climate change and urbanization—acting in concert with the global spread of invasive species—has set the stage for spontaneous vegetation to play a major ecological role in the human-dominated landscapes of the future (Ziska et al. 2004; Christopher 2008; George et al. 2009). Regardless of how humans feel about this brave new ecology, the plants described in this book are well adapted to the world we have created and, as such, are neither good nor bad—they are us.

Landscaping with Spontaneous Urban Vegetation

Most people tend to interpret the presence of spontaneous urban vegetation in their neighborhood as a visible manifestation of dereliction and neglect while viewing the same plants growing in a suburban or rural context as "wildflowers" (think about the beautiful combination of chicory and Queen Anne's lace along the road-side in July and August). Clearly, the context in which a plant exists has everything to do with how people feel about it. The very same species that people despise as invasive in North America are often cherished natives in Europe. The common reed, for example, is widely distributed in river deltas throughout Europe and serves as a major bulwark against costal erosion. The fact that the plant is dying throughout much of its native European range is viewed as an ecological crisis. The same species growing in the same type of habitat in eastern North America, however, is widely seen as a noxious weed that should be eradicated.

This relativity is explained by the fact that people are looking at the same plant through the subjective lens of a cultural value judgment which places a higher value on the form or nativity of a given plant association than on its function. While this dichotomy may be appropriate and necessary for preserving wilderness areas or rare habitats, it does not work so well in the urban context, where ecological functionality should be recognized as being of equal value to ecological form (Sagoff 2005). Unfortunately, the aesthetics of ecologically functional, spontaneous urban landscapes often leave something to be desired, which raises the question of whether

In August, Queen Anne's lace (*Daucus carota*) and chicory (*Cichorium intybus*) make a stunning combination along roads in the Northeast.

A typical spontaneous urban meadow in which purple loosestrife (*Lythrum salicaria*), reed canarygrass (*Phalaris arundinacea*), chicory (*Cichorium intybus*), mullein (*Verbascum thapsus*), evening primrose (*Oenothera biennis*), quackgrass (*Elymus repens*), and Queen Anne's lace (*Daucus carota*) are clearly identifiable.

or not there is a way to harmonize such spontaneous ecological functionality with people's desire to live in a neat and tidy environment.

As is the case with urban ecology in general, the Europeans have been hard at work on this question for the past 20 to 30 years. *The Dynamic Landscape*, edited by Nigel Dunnett and James Hitchmough introduces North American readers to years of European work on ways to manipulate the aesthetic characteristics of spontaneous vegetation through the judicious addition or deletion of species. The book is unique in combining solid ecological research on natural grasslands and agricultural meadows with the horticultural goal of creating aesthetically pleasing, low-maintenance landscapes. Indeed, landscapes that include spontaneous vegetation fit the technical definition of *sustainable* in the sense that they are adapted to the site, require minimal maintenance, and are ecologically functional (Kühn 2006).

Using the work of various European authors as a guide, I have developed the concept of the "cosmopolitan urban meadow" as a way of capitalizing on the aesthetic and ecological opportunities presented by some of the herbaceous species covered in this book. The plants chosen for inclusion in this novel landscape form (Appendix 4) were selected on the basis of their ability to meet the following criteria:

- They should be long-lived and capable of spreading vegetatively (**erosion control value**).
- They should be tolerant of full sun, drought, soil compaction, road salt, and high or low soil pH (**stress tolerance**).
- They should not be too tall or weedy looking and should have some obvious ornamental characteristics (**aesthetic value**).
- They should be attractive to pollinating and herbivorous invertebrates and seed-eating animals (**wildlife value**).
- They should be commercially available as seed (**economic value**).

The basic idea behind the cosmopolitan urban meadow is to select an assemblage of plants that will grow well on typical urban soil, create an aesthetically pleasing urban meadow on vacant land, and remain in place until a more permanent use for the land is developed. Some minimal soil preparation would be required to get the meadow established, but much less than it would take to start a lawn from seed. There would also have to be some selective weeding to keep out unsightly plants such as ragweed and mugwort, especially in the early stages of growth. An annual mowing in late summer or fall would maintain the composition of the meadow and keep woody plants from taking over.

The question of aesthetics is more complicated for landscapes dominated by spontaneous woody plants, which have longer life spans and less conspicuous ornamental attributes than herbaceous species. Nevertheless, the appearance of an urban woodland can often be enhanced by removing trees or shrubs that are unsightly, in poor health, or too crowded. Such landscape "editing," if done by a trained horticulturist, can dramatically improve both the appearance and the recreation potential of a spontaneous urban woodland (Kowarik and Langer 2005). The foundation for this approach to vegetation management was laid out in 1966 by Frank Egler in *The Wild Gardener in the Wild Landscape,* a remarkable book that he wrote under the pseudonym Warren G. Kenfield. Egler was light-years ahead of his time in advocating the use of herbicides to remove unwanted trees in order to arrest or suspend forest succession at the shrub or grassland stage of development. He called this process "intaglio," after the engraving technique that creates an image by removing unwanted material from the surface of a blank plate.

Another landscape concept that utilizes spontaneous vegetation is the "freedom lawn" introduced in the landmark book *Redesigning the American Lawn* (Bormann, Balomori, and Geballe 1993). The freedom lawn, as the authors define it, is a collection of low-growing, spontaneous plants that "results from an interaction of

A stand of Norway maples (*Acer platanoides*) dominates a typical highway slope in Watertown, Massachusetts.

naturally occurring processes and the selective effects of lawn mowing." The book goes on to contrast the freedom lawn with the "industrial lawn," which requires the constant input of chemical fertilizers, weed killers, and water in order to maintain a uniform monoculture of grass. The beauty of the freedom lawn concept is that it has the potential to transform a wasteful sink of petroleum products into a sustainable source of ecological benefits.

And finally there is the concept of the "spontaneous roof," a modification of the increasingly popular "green roof." Most current designs for the so-called extensive green roof involve the use of various types of succulent stonecrops in the genus *Sedum* growing in a lightweight medium consisting of a high percentage of expanded slate or shale on top of several layers of waterproofing. Sedums are low growing and colorful, and once established require very little maintenance. The only problem is that they grow very slowly and sometimes have a difficult time holding the ground against other, more aggressive species. An alternative strategy is the "brown roof" described in *Planting Green Roofs and Living Walls* (Dunnett and Kingsbury 2004). According to the authors, brown roofs "have been covered with substrate or loose material but have not been purposefully planted. . . . Brown roofs are created primarily for biodiversity purposes and aim to recreate typical brownfield conditions through the use of by-products of the development of urban sites: brick rubble, crushed concrete, and subsoils. Such roofs may colonize spontaneously with vegetation but the unvegetated loose substrates can also provide habitat for a range of invertebrates and birds." If one tweaks this definition by adding a somewhat richer substrate and managing the vegetation by selectively removing woody plants and tall-growing or toxic herbaceous perennials, the brown roof concept can be transformed into what I call a spontaneous roof. Indeed, in the absence of ongoing maintenance, it seems likely that many of the extensive green roofs being built today are destined to become tomorrow's spontaneous roofs.

How to Use This Book

Following the format of *Weeds of the Northeast*, also published by Cornell University Press, I have arranged the plants in this book according to major botanical categories. The highest level is the Taxonomic Group, which for the purposes of this book includes *ferns, horsetails, conifers, woody dicots, herbaceous dicots*, and *monocots* (see Glossary for definitions). Within each of these groupings are the plant families, which are arranged alphabetically; within each family the individual species are listed alphabetically. Appendix 5 describes the key characteristics of 12 common plant families that account for 64% of the species covered in this book. Being able to recognize the higher-order characteristics of a given plant family is a critical step not only in determining the identity of a specific species but also in getting a handle on the daunting diversity of the plant kingdom.

While determining the correct Scientific Name of a plant may seem to be a straightforward process, it is anything but that. Modern molecular technology has led to a revolution in plant taxonomy, which in turn has led to a host of new

FORMAT USED FOR THE PLANT ENTRIES*

Scientific Name and Author   Common Name

SYNONYMS: older scientific names and alternative common names

LIFE FORM: annual, biennial, herbaceous perennial, or woody (including shrubs, trees, and vines), and the typical height in urban conditions

PLACE OF ORIGIN: where the plant grows as a native species

VEGETATIVE CHARACTERISTICS: a description of the vegetative features of the mature plant (leaves, stems, branches, trunk, bark, and root system)

FLOWERS AND FRUIT: a description of the flowers, fruit, and seeds, as well as the timing of these events

GERMINATION AND REGENERATION: the conditions required to stimulate seed germination and, for perennials, the mode of vegetative sprouting from underground structures or woody stems

HABITAT PREFERENCES: the types of urban habitats where the plant is most likely to be found

ECOLOGICAL FUNCTIONS: the "ecological services" the plant contributes to the urban environment

CULTURAL SIGNIFICANCE: the various uses—both ancient and modern—of the plant and a few details about the history of its introduction into the Northeast

RELATED SPECIES: names and brief descriptions of closely related plants that are typically not as common as the primary entry

SIMILAR SPECIES: how to distinguish the primary species from other species treated in the book

HIGHER TAXONOMIC GROUP: *ferns, horsetails, conifers, woody dicots, herbaceous dicots, or monocots* followed by the **Plant Family**

* Plant entries may not include all categories.

names for species whose names had remained unchanged for a hundred years. It is a certainty that some of the names of the species treated in this book will have been changed during the interval between when the manuscript for this book was finalized for publication and when it was printed. Equally problematic is the issue of a plant's COMMON NAME, which frequently varies regionally. To help resolve this potentially confusing situation I used the standardized weed names—both scientific and common—published in April 2007 on the website of the Weed Science Society of America (http://www.wssa.net/Weeds/ID/WeedNames/namesearch.php). For plants not treated in this source I used the Plants Database of the U.S. Department of Agriculture (http://plants.usda.gov/). Each species account lists SYNONYMS—for both scientific and common names—that have been used for the plant over the past hundred years or so.

The PLACE OF ORIGIN of a given species is not always easy to pin down, either, especially for plants with a long history of association with humans. I have had to rely on a number of different sources, the most important of which are the *Manual of Vascular Plants of Northeastern United States and Adjacent Canada* (2nd edition, 1991) by H. A. Gleason and A. Cronquist, and *Weeds of Lawn and Garden* by John Fogg Jr. (first published in 1945). I particularly admire the latter book not only for its comprehensive treatment of species and accessible text but also for the beautiful and informative line drawings by Léonie Hagerty.

I have tried to keep the technical descriptions of the plants' physical characteristics simple and jargon-free, a goal that is possible only because of the many photographs that accompany the text. The Glossary defines all the technical terms used in the text. Under the subheads HABITAT PREFERENCES, ECOLOGICAL FUNCTIONS, and CULTURAL SIGNIFICANCE I explain where the plant grows, the ecological functions it serves in the urban environment, and its relationship with humans.

I have not included information on how to kill the described plant. Virtually every other book on weeds and invasive species already provides more than enough information on this topic. Indeed, these books' authors seem to assume that the main reason for learning to identify weeds is to be able to kill them more effectively. My decision is in keeping with this book's two primary goals: to teach people how to identify the plants that are growing in urban areas, and to counter the widespread perception that these plants are ecologically harmful or useless and should be eliminated from the landscape. This is not to say that I think these species should be actively cultivated, but I do hope that people will develop a sense of respect and appreciation for what these plants are contributing—free of charge—to the quality of urban life.

A Note on the Photographs

Except where noted in the captions, all photographs in the book were taken outdoors by the author with a digital camera (either an Olympus 8080 or a Canon Digital Rebel XTi with a 60 mm macro lens). The photos were selected for their ability to capture the essence of the plant growing in its natural habitat, with an emphasis on important macroscopic features that can be used to identify the plant at various stages in its life cycle. The extensive use of photographs has allowed me to keep technical descriptions to an absolute minimum and allows an untrained observer confronted with an unknown plant to ignore the formal classification system and simply "go fishing" through the pictures for a match.

Onoclea sensibilis L.   Sensitive Fern

SYNONYM: bead fern

LIFE FORM: **herbaceous perennial**; up to 2 feet (60 cm) tall

PLACE OF ORIGIN: eastern North America

VEGETATIVE CHARACTERISTICS: The tall, leathery leaves or fronds are roughly triangular in outline and up to 2 feet (60 cm) long; they are dissected into 12 or more opposite pairs of leaflets and those at the base are more lobed than those near the apex. Young, expanding leaves are pale red; mature leaves are light green with scattered white hairs on the underside. The foliage dies down to the ground with the first frost (hence the common name). A forking brown rhizome, which grows just below the surface, gives rise to new shoots and the fibrous root system.

REPRODUCTIVE STRUCTURES: Sensitive fern reproduces by wind-dispersed spores that are produced on stiff fertile fronds that stand about 1 foot (30 cm) tall. These fronds have numerous short branches that are lined with small, bead-like capsules that produce the spores; they are green during the growing season and turn dark brown at maturity. They typically remain standing through the winter.

GERMINATION AND REGENERATION: The spores germinate on moist soil and develop into tiny gametophytes that give rise to the sporophyte generation that we recognize as a fern. Once established, sensitive fern spreads by means of branching rhizomes, eventually forming large clumps.

HABITAT PREFERENCES: Sensitive fern is more sun tolerant than other ferns and reproduces freely under urban conditions. It grows best in sunny, moist sites but can also be found in dry, shady situations. It is common along the margins of freshwater wetlands, ponds, streams, and swamps; drainage ditches; woodland understories; urban meadows; and railroad rights-of-way.

ECOLOGICAL FUNCTION: Disturbance-adapted colonizer of bare ground.

RELATED SPECIES: **Hayscented fern (*Dennstaedtia punctilobula* (Michx.) Moore) (Family: Dennstaedtiaceae)**, a North American native, is very different in appearance but is also common in the urban environment, especially in dry, shady woodlands. Its leaves are about 16 inches (40 cm) long and are divided into 20 or so pairs of leaflets that are in turn divided into subleaflets that bear small "spore dots" on their undersides. Under the right conditions hayscented fern can spread rapidly and form large persistent clumps; it dies back to a perennial rootstock with the first frost. Hayscented fern seems to be increasingly abundant in forest understories; probably because of its tolerance of the low soil pHs caused by acid precipitation.

Emerging leaves of sensitive fern growing at the edge of a pond

Sensitive fern frond

Dry fertile fronds of sensitive fern

Dense stand of sensitive fern

Hayscented fern growing in a moist microclimate

Hayscented fern fronds

Equisetum arvense L.   Field Horsetail

SYNONYMS: common horsetail, horsetail fern, bottle brush, jointed rush

LIFE FORM: **herbaceous perennial**; up to 1.5 feet (45 cm) tall

PLACE OF ORIGIN: Europe, Asia, and North America

VEGETATIVE CHARACTERISTICS: The stiff, hollow green stems emerge in late May; as the season progresses they produce whorls of secondary branches that give them a "bottle-brush" appearance. The scalelike true leaves, which are located at the base of the secondary branches, are inconspicuous. The stems are rough to the touch and die back to the ground in late fall.

REPRODUCTIVE STRUCTURES: In early spring, field horsetail will occasionally produce unbranched fertile stems that may be up to 1 foot (30 cm) tall. These stems are brownish to pale pink and are topped with a small cone, the strobilus, which releases thousands of microscopic spores before withering away. Horsetails are considered to be the most primitive of the fern-type plants that produce spores rather than seeds.

GERMINATION AND REGENERATION: Field horsetail reproduces from wind-dispersed spores; established plants can form large clumps that increase in size by means of deep rhizomes.

HABITAT PREFERENCES: This species is commonly found in sandy soil along roadsides as well as in areas with poor drainage. It can form large clumps along highway banks covered with coarse gravel and among the ballast supporting railroad tracks because it is resistant to many of the herbicides used to kill flowering plants.

ECOLOGICAL FUNCTIONS: Disturbance-adapted colonizer of bare ground; erosion control on slopes.

CULTURAL SIGNIFICANCE: Native Americans ate the young shoots of field horsetail and used a tea made from the stems to treat kidney and bladder disorders. European traditional medicine used an extract of the stems as a diuretic and an astringent. Field horsetail accumulates silica in its tissues—up to 12% of its dry weight—and also accumulates various heavy metals, including selenium, gold, and mercury, which suggests that it may be useful for cleaning up contaminated soil (phytoremediation). Extinct horsetail species were abundant during the Carboniferous period (250 million years ago); much of the coal we burn today is composed of their compressed remains.

RELATED SPECIES: Scouring rush (*Equisetum hyemale* L.) consists entirely of green stems, lacking the whorled secondary branches of field horsetail. While scouring rush is common along railroad tracks (because of its resistance to herbicides), it is more typically found in shady, moist areas. The silica deposits on the stiff stems of both species have led to their use as "organic" scouring pads for scrubbing pots and pans on camping trips.

Field horsetail
vegetative stem

Field horsetail growing on a gravel-covered highway embankment

Field horsetail
spore-producing
stems (photo by
Les Mehrhoff)

Scouring rush
growth habit

Scouring rush growing in railroad ballast

Scouring rush spore-
producing cone

Taxus cuspidata Sieb. & Zucc.   Japanese Yew

Synonym: spreading yew

Life Form: evergreen tree; 10–50 feet (3–15 m) tall

Place of Origin: northeast Asia

Vegetative Characteristics: Cultivated Japanese yews, grown from cuttings, have a spreading, shrubby form with dense foliage; spontaneous trees that grow from seed tend to develop a single central trunk and have sparse foliage. The dark, evergreen needles are arranged alternately along the stem; they are short stalked and 0.5–1 inch (1–2.5 cm) long. The distinctive reddish brown bark peels off in thin strips.

Reproductive Structures: All yews are dioecious, with separate male and female individuals. On both sexes the cones develop in late fall in the axils of the needles and go through the winter in a rudimentary state of development. Male cones are produced in abundance and are clearly visible in fall along the underside of the year-old shoots. Female plants produce many fewer cones than male plants. Male cones shed their wind-dispersed pollen in late spring, and female plants produce bright red "fruits" in the fall that birds readily consume and disperse. The fruit consists of a hard brown seed (which is highly toxic) surrounded by a fleshy red aril (which is nontoxic).

Germination and Regeneration: Seeds germinate in spring; seedlings grow best in light to dense shade and soil rich in humus.

Habitat Preferences: Since the 1980s Japanese yews have started to appear with greater frequency in the understory of disturbed or emergent urban forests; they grow slowly and can persist in a suppressed state for many years.

Ecological Function: Food and habitat for wildlife.

Cultural Significance: The cultivar 'Nana', introduced as an ornamental into North America in the late 1800s, has become a ubiquitous hedge and foundation plant in the Northeast because of its low, spreading growth habit and because it is hardier than the English yew (*T. baccata*). In the early 1980s the anti–ovarian cancer drug taxol was isolated from the bark of the Pacific yew (*T. brevifolia*); subsequent research has determined that taxol occurs in the foliage of all *Taxus* species.

Japanese yew growing spontaneously in the understory of a Norway maple woodland

Japanese yew foliage

Japanese yew seedling

Japanese yew hedge in the Boston area

Male cones of Japanese yew along the undersides of 1-year-old shoots

Mature female cones of Japanese yew

Acer negundo L.　Box Elder

SYNONYMS: ash-leaved maple, Manitoba maple, maple ash

LIFE FORM: deciduous tree; up to 60 feet (18 m) tall

PLACE OF ORIGIN: North America

VEGETATIVE CHARACTERISTICS: Box elder typically produces a short, leaning trunk or multiple trunks with a broad, spreading crown. The opposite, pinnately compound leaves consist of 3–7 coarsely toothed leaflets, 2–4 inches (5–10 cm) long, that are extremely variable in shape. The young twigs are purplish or green and are often covered with a white, waxy coating that rubs off easily. The leaves turn pale yellow in the fall.

FLOWERS AND FRUIT: Box elder produces conspicuous chains (racemes) of green wind-pollinated flowers, 4–6 inches (10–15 cm) long, in late spring. These are followed by strings of winged samaras that persist through most of the winter, making the tree easy to identify in the leafless condition. Box elder produces unisexual flowers on separate male and female (seed-bearing) trees.

GERMINATION AND REGENERATION: Seeds are wind dispersed from fall through winter and germinate on moist, bare ground in spring. Saplings and mature trees sprout readily from the base following injury or partial uprooting of the trunk, producing a multistemmed growth form.

HABITAT PREFERENCES: This disturbance-adapted species typically grows along streams and rivers in full sun. In the urban environment it grows in a wide variety of sites including the margins of freshwater wetlands, vacant lots, railroad rights-of-way, small pavement openings, rock outcrops, and along chain-link fence lines. It is a remarkably hardy and adaptable tree with a widespread distribution across most of Canada, the United States, and into Mexico and Guatemala.

ECOLOGICAL FUNCTIONS: Food and habitat for wildlife; stream and river bank stabilization; soil building on degraded land; heat reduction in paved areas; tolerant of roadway salt and compacted soil.

CULTURAL SIGNIFICANCE: Native Americans used a tea made from the bark to induce vomiting, and the sap in spring can be boiled down to make syrup. Horticultural varieties of box elder with variegated foliage have limited availability in the nursery trade. The common name of the plant derives from the resemblance of its wood to that of boxwood (genus *Buxus*) and of its foliage to that of the elderberry bush (genus *Sambucus*).

Box elder
growth habit

Box elder seedling with juvenile foliage

Compound leaves of box elder

Box elder resprouting
vigorously after being cut back

Multistemmed growth
form of a cut-back plant,
note the silvery white stems

Box elder seeds

Acer platanoides L. Norway Maple

SYNONYMS: none

LIFE FORM: deciduous tree; 40–60 feet tall (12–18 m)

PLACE OF ORIGIN: northern and central Europe, western Asia

VEGETATIVE CHARACTERISTICS: Norway maple develops a broad, spreading crown; its bark is smooth and gray when young and becomes furrowed with age. The opposite, dark green leaves are 4–7 inches (10–18 cm) wide and have the classical maple shape—with 5 main lobes; they turn yellow in late fall, long after the native maples have dropped their leaves. The petiole exudes milky sap when broken, a diagnostic feature that distinguishes this species from the native sugar maple (*Acer saccharum*).

FLOWERS AND FRUIT: Norway maple produces clusters of distinctive chartreuse wind-pollinated flowers in early spring before the leaves emerge; these are followed by paired samaras with wings that spread apart at an angle of 180°. The head of the seed containing the embryo is relatively flat compared with that of other maples.

GERMINATION AND REGENERATION: The wind-dispersed seeds mature in the fall and can travel 100 feet (30 m) or more from their parent tree. Seeds will germinate in either sun or shade, and saplings resprout readily from the base following damage.

HABITAT PREFERENCES: Because it tolerates shade and compacted soils, Norway maple has become a dominant late successional tree in many disturbed urban woodlands. It is also common along streams and riverbanks, in neglected residential and commercial landscapes, in vacant lots and dumps, and along chain-link fence lines.

ECOLOGICAL FUNCTIONS: Tolerant of roadway salt and compacted soil; heat reduction in paved areas; erosion control on slopes.

CULTURAL SIGNIFICANCE: Norway maple is among the most commonly planted street trees in the Northeast. Numerous horticultural selections are available, including purple-leaved ('Crimson King') and fastigiated cultivars. Norway maple was introduced into North America in 1756 by John Bartram of Philadelphia and became popular during the late 1800s. It was extensively planted in the 1950s and 1960s to replace American elms wiped out by Dutch elm disease. Many states list it as an invasive species.

Norway maple leaves turn butter yellow in late fall

Chartreuse
Norway
maple
flowers in
early spring

Milky sap
distinguishes
Norway maple
from other
maples in this
book

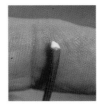

Developing seeds of Norway maple

Typical Norway maple leaf

Mature Norway maple
bark

Acer pseudoplatanus L.   Sycamore Maple

SYNONYM: planetree maple

LIFE FORM: **deciduous tree**; up to 80 feet tall (25 m)

PLACE OF ORIGIN: Europe and western Asia

VEGETATIVE CHARACTERISTICS: Sycamore maple produces coarsely toothed, opposite leaves that are 3–6 inches (7–15 cm) long and wide, and often have red petioles; the smooth, 5-lobed leaves have a leathery texture and are dark green above and greenish white beneath. The bark on mature trees can vary from pale gray to reddish brown and flakes off in large, irregular patches. The leaves turn pale yellow in fall.

FLOWERS AND FRUIT: Sycamore maple produces small yellow-green, wind-pollinated flowers in May on pendulous chains (racemes) that are 2–6 inches (5–15 cm) long. These are followed by chains of winged seeds (samaras) that mature in the fall. The seeds are produced in pairs and have broad wings that form a 60–90° angle.

GERMINATION AND REGENERATION: Sycamore maple seeds are dispersed by wind in autumn, and seedlings germinate in spring under a wide variety of conditions from full sun to part shade. The tree is quite tolerant of salt spray and dry soil.

HABITAT PREFERENCES: Because of its salt tolerance, sycamore maple was widely planted as an ornamental in coastal landscapes along the U.S. East Coast. It has now fallen out of favor with landscapers, but mature specimens persist and seedlings grow well in a variety of disturbed urban sites and along roadways treated with salt.

ECOLOGICAL FUNCTIONS: Tolerant of salt along roadways and near the ocean; heat reduction in paved areas; erosion control on slopes; tolerant of compacted soil; food and habitat for wildlife.

CULTURAL SIGNIFICANCE: There are numerous horticultural selections of this species, most with brightly colored or variegated foliage.

RELATED SPECIES: The leaves of **Norway maple** (*Acer platanoides*) and **sugar maple** (*Acer saccharum*) lack fine teeth along their margins, and those of **red maple** (*Acer rubrum*) are smaller and lighter green.

Sycamore maple foliage and seeds

Sycamore maple growth habit

Sycamore maple seedlings often grow where blacktop meets a building

Leaves of a vigorous sycamore maple sprout

The mature bark of sycamore maple

Sycamore maple inflorescence (photo by Les Mehrhoff)

Acer saccharinum L.   Silver Maple

Synonyms: *Acer dasycarpum*, soft maple

Life Form: deciduous tree; up to 100 feet (30 m) tall

Place of origin: eastern North America

Vegetative Characteristics: Silver maple is one of the tallest trees native to the Northeast. It has light gray bark and a broad, open crown, and often produces multiple trunks. The opposite leaves are about 6 inches (15 cm) long and typically have 5 lobes; the margin of the leaf is coarsely serrated and the sinuses separating the lobes extend nearly to the central vein; the upper surface is light green and the underside silvery white; fall color is pale yellow.

Flowers and Fruit: While not strictly a dioecious species, individual silver maples (as well as most other maples) produce flowers that are predominantly male or female, so it usually takes more than one tree to produce viable seed. Silver maple is among the first of our native trees to bloom, producing small, reddish, wind-pollinated flowers from mid-March through early April. The female flowers quickly mature into broad-winged samaras up to 2 inches (5 cm) long.

Germination and Regeneration: The seeds are dispersed by the wind in late spring, before the leaves have fully expanded; they germinate without delay on exposed, sunny sites. Established trees sprout readily from the base when injured or partially uprooted by stream flow, resulting in a multistemmed growth form.

Habitat Preferences: Silver maple grows best in sunny, moist habitats, especially freshwater wetlands and along the banks of ponds, streams, and rivers.

Ecological Functions: Heat reduction in paved areas; food and habitat for wildlife; stream and river bank stabilization; nutrient absorption in wetlands; tolerant of salt and compacted soil.

Cultural Significance: Silver maples were once widely planted as street trees because of their rapid growth, but this same factor has led to problems because they often interfere with overhead wires and create cracks in sidewalks. The species also has a reputation for being weak wooded and dropping large branches, but this is usually the result of poor pruning done to compensate for putting the tree in the wrong place.

Related Species: Red maple (*Acer rubrum* L.) is native to eastern North America and can reach heights up to 100 feet (30 m). While red maple can grow in a variety of habitats, it is most common in bottomland situations or sites with good soil moisture. It produces bright red flowers in very early spring on separate male and female plants, and disperses its small—about 0.75 inch (2 cm) long—winged seeds in late spring. The leaves are opposite and typically have 3, but often 5, lobes with serrated margins; they turn bright red or yellow in the fall. The bark of young trees is smooth and light gray, but it becomes rough and scaly with age. Horticultural selections of red maple are widely cultivated as street trees, primarily for their bright red fall foliage.

A mature silver maple growing along the Charles River in Cambridge, Massachusetts

Female flowers of silver maple in early spring

Silver maple foliage has silvery-white undersides

Silver maple seeds (*right*) and red maple seeds (*left*) are shed in mid- to late spring

Red maple leaves are bright red in fall

Red maple seeds begin to develop before the leaves come out in spring

Rhus typhina L.   Staghorn Sumac

SYNONYMS: *Rhus hirta,* velvet sumac, vinegar tree

LIFE FORM: **multistemmed shrub**; up to 30 feet (9 m) tall

PLACE OF ORIGIN: eastern North America

VEGETATIVE CHARACTERISTICS: Staghorn sumac produces sparsely branched, crooked stems—often looking like antlers—and 1-year-old twigs that are densely covered with velvety hairs (hence the common name). The alternate, pinnately compound leaves are composed of 5–15 pairs of leaflets with serrated margins and are 1–2 feet (30–60 cm) long. In October the leaves turn spectacular shades of red, orange, and yellow.

FLOWERS AND FRUIT: Staghorn sumac produces terminal panicles of greenish white, insect-pollinated flowers in late spring on separate male and female plants. Cone-shaped clusters of deep red, velvety fruits, up to 6 inches (15 cm) long, terminate the branches of female plants in late summer and persist through the winter.

GERMINATION AND REGENERATION: Seeds are consumed and dispersed by numerous species of birds and germinate in sunny, open sites; established plants produce root suckers and form large, dense thickets.

HABITAT PREFERENCES: Staghorn sumac grows best in full sun and well-drained soil. In the urban environment it is common in minimally maintained public parks, vacant lots, rubble dumps, degraded or emergent woodlands, rock outcrops and stone walls, unmowed highway banks, and railroad rights-of-way.

ECOLOGICAL FUNCTIONS: Tolerant of roadway salt and compacted soil; food and habitat for wildlife; erosion control on slopes.

CULTURAL SIGNIFICANCE: Juice from the crushed berries, when steeped in water, can be used as a gargle for sore throat or as a refreshing, and somewhat acidic, summer tonic (the hairs on the surface of the seeds contain malic acid). Native Americans used the leaves and fruits as a poultice to soothe irritated skin. Staghorn sumac is widely planted as a native erosion-control species.

RELATED SPECIES: Smooth sumac (*Rhus glabra* L.) is somewhat smaller than staghorn sumac (up to 24 feet [7 m] tall) and lacks velvety hairs on its purplish stems and petioles. It sprouts vigorously from the roots, forming large clumps, and its alternate leaves turn brilliant red in the fall. William Darlington (1859) noted that "this shrub is apt to be abundant in neglected sterile old fields; and its prevalence, in arable lands, is strong evidence of the occupant being a poor thriftless farmer." Native Americans and European immigrants used extracts of the bark, leaves, and fruit to treat asthma and variety of other ailments, as well as for tanning leather and making dyes.

Staghorn sumac foliage and fruits

New staghorn sumac stems develop from root sprouts and can form large colonies

Smooth sumac's outstanding fall color

Staghorn sumac fruits ready for dispersal

Distinctive hairy stems of staghorn sumac

Hairless, light purple, smooth sumac stems

Toxicodendron radicans (L.) Kuntze   Poison Ivy

SYNONYMS: *Rhus radicans, Rhus toxicodendron,* poison vine

LIFE FORM: deciduous vine; up to 50 feet (15 m) long

PLACE OF ORIGIN: eastern North America

VEGETATIVE CHARACTERISTICS: This ubiquitous and highly variable vine can climb tall trees, grow as a ground cover, or form a dense, spreading shrub. Regardless of its growth habit, all of its branches have a distinct horizontal orientation. The older climbing stems produce conspicuous aerial roots (the meaning of the word *radicans*) that give them a "bearded" appearance. The alternate, compound leaves are composed of 3 glossy leaflets (source of the old warning "leaves of three, let them be"), each about 4 inches (10 cm) long with smooth or coarsely toothed margins; the terminal leaflet always has a petiole while the lateral leaflets are nearly sessile. Leaves of plants growing in the shade turn dull yellow in fall; plants in full sun turn bright red.

FLOWERS AND FRUIT: Poison ivy produces clusters of inconspicuous 5-petaled, yellowish green, insect-pollinated flowers in the axils of the leaves from May through June on separate male and female plants. The berries on female plants turn from green to gray to white when they mature in September.

GERMINATION AND REGENERATION: Fruits are eaten by birds, and seeds germinate beneath their roosts. Established plants spread by underground rhizomes and stems that root where they touch the ground.

HABITAT PREFERENCES: Poison ivy grows best in moist soils in shade or full sun, but can also be found in dry, sandy sites. Its tolerance of high salt concentrations accounts for its abundance near the ocean as well as along busy roadsides. In the urban environment it is common on rock outcrops and stone walls; climbing up telephone poles, buildings, and chain-link fences; along unmowed highway banks and railroad tracks; and climbing up tree trunks in moist or dry woodlands.

ECOLOGICAL FUNCTIONS: Tolerant of roadway salt and compacted soil; food and habitat for wildlife; erosion control on slopes.

CULTURAL SIGNIFICANCE: Touching any part of this plant, in summer or winter, causes allergenic dermatitis in 60–80% of people. The offending ingredient is urushiol, which is located in the sap. Once absorbed through the skin it causes a characteristic itchy rash within a day or two of contact. Poison ivy has been used in traditional medicine and was one of the herbs sold by the Shakers to treat chronic paralysis, rheumatism, skin diseases, and bladder paralysis. In 1624 Captain John Smith became the first European to describe the plant, which "being but touched causeth rednesse, itching, and lastly blisters, the which howsoever after a while passe away of themselves without further harm."

SIMILAR SPECIES: Poison ivy is often confused with Virginia creeper (*Parthenocissus quinquefolia*), a climbing vine with 5 leaflets per leaf and nonhairy stems.

Aerial roots give poison ivy stems a distinctive "bearded" appearance

Poison ivy foliage

Poison ivy will climb on anything

The mature, horizontal growth habit of poison ivy growing on an iron fence

Poison ivy fruits ripen in September

Poison ivy in full fall color along a roadside chain-link fence

Berberis thunbergii DC   Japanese Barberry

SYNONYM: Thunberg's barberry

LIFE FORM: **deciduous shrub**; 2–6 feet (0.6–2 m) tall

PLACE OF ORIGIN: Japan

VEGETATIVE CHARACTERISTICS: The stems of Japanese barberry are grooved and brown; the alternate leaves are about 0.5 inch (2 cm) long and spoon or egg shaped with smooth edges. An unbranched, sharp spine at each node makes the whole plant painful to touch. The twigs, when broken, are bright yellow, as are the roots and rhizomes. Japanese barberry is one of the first shrubs to leaf out in early spring; its fall color is variable, ranging from yellow to orange to scarlet.

FLOWERS AND FRUIT: The small, yellow, cup-shaped flowers have 6 petals and are produced in April and May; they line the length of the stem, hanging down from leaf axils in small clusters and are pollinated by insects. The showy red fruits, which are about 0.5 inch (1 cm) long, mature in the fall and persist into early winter. Plants growing in the sun produce much more fruit than those in the shade.

GERMINATION AND REGENERATION: Seeds are dispersed by birds and germinate in sunny or semishady conditions. Stems root where they touch the ground; established plants spread by underground rhizomes.

HABITAT PREFERENCES: Japanese barberry is a highly adaptable shrub that grows as well in dry soil and full sun as in moist shade. Because of this adaptability it has been widely planted in low-maintenance plantings such as parking lot islands and municipal and commercial landscapes, where it persists long after most other plants have died. In the woodland understory it conspicuously leafs out in spring before most native species and in the fall retains its green leaves longer than the surrounding plants. Deer generally avoid eating the foliage or twigs of Japanese barberry.

ECOLOGICAL FUNCTIONS: Tolerant of roadway salt and compacted soil; food and habitat for wildlife; erosion control on slopes.

CULTURAL SIGNIFICANCE: This Japanese species was introduced into North America as an ornamental in 1875 from Japan via St. Petersburg, Russia, to Boston. By the early 1900s there were reports of the plant spreading on its own. Cultivars with colored foliage, especially purple, have been widely planted. The fruit can be used to make jelly, and the bright yellow roots, which contain the medicinally active compound berberine, have been used as a digestive tonic and a purgative. Many states list Japanese barberry as an invasive species because of its ability to reproduce in the forest understory.

Japanese barberry leafs out in spring, well before most native woody plants

Japanese barberry flowers and foliage in spring (photo by Les Mehrhoff)

Japanese barberry growing at the edge of the woods

Japanese barberry produces fall color in early November

Japanese barberry produces fruits in the autumn

Alnus glutinosa (L.) Gaertner Black Alder

Synonyms: *Alnus vulgaris*, European alder

Life Form: deciduous tree; 35–60 feet (10–18 m) tall

Place of Origin: Europe, western Asia, and North Africa

Vegetative Characteristics: Black alder produces shiny, alternate leaves that are round or oval with wavy margins, prominent veins, and a notch at the tip. They are 2–4 inches (5–10 cm) long and nearly as wide, and typically stay green late into the autumn, when they fall without changing color. The bark of the mature tree is dark brown or black with flaking plates and vertical fissures.

Flowers and Fruit: Separate male and female flowers are produced on the same plant (monoecious). The pendulous male catkins are conspicuous in early spring when they elongate and shed their pollen to the wind; the woody female "cones" persist on the trees through winter until early spring and help to identify this species.

Germination and Regeneration: Seeds are dispersed by water or wind and germinate in the spring on moist, bare ground in full sun; established trees sprout readily from the base following injury. In Europe trees are typically cut down to the ground on a 10-year cycle (coppice forestry) to produce poles and small-diameter fuel wood.

Habitat Preferences: Black alder is common along stream and river banks and can form dense thickets in freshwater wetlands.

Ecological Functions: Soil improvement; tolerant of road salt and saturated soil; food and habitat for wildlife; stream and river bank stabilization.

Cultural Significance: Black alder was originally introduced for ornamental, forestry, and soil stabilization purposes. It also improves soil quality by fixing atmospheric nitrogen in symbiosis with an actinobacteria in the genus *Frankia*. In Europe the bark and the young shoots have long been used as dyes, and the leaves for tanning leather. It also had limited use as a medicinal plant to treat swellings and inflammation, and its rot-resistant wood has been used for pilings.

Male (pendulous) and female (upright) black alder flowers

Black alder keeps its foliage longer than most native trees

Male flowers of black alder are conspicuous in very early spring

Black alder foliage

Spent female cones (fruits) of black alder from the previous year (*left*) along with the current season's cones (*right*)

Betula nigra L.   River Birch

SYNONYM: red birch

LIFE FORM: deciduous tree; 30–60 feet (10–18 m) tall

PLACE OF ORIGIN: eastern North America

VEGETATIVE CHARACTERISTICS: When young, river birch produces freely peeling, buff-colored bark; as the tree ages, the bark become dark reddish brown or black with deep furrows and breaks off in irregular plates. Under natural conditions the tree typically develops a multistemmed growth habit. The alternate, simple leaves—up to 3.5 inches (9 cm) long by 2.5 inches (6 cm) wide—are roughly diamond arrowhead shaped, sharp pointed, and have serrated margins. The leaves turn yellow in the fall.

FLOWERS AND FRUIT: River birch produces separate male and female flowers on the same plant (monoecious); the pendulous male catkins are conspicuous in early spring when they expand to a length of 2–3 inches (5–7 cm) and shed their pollen to the wind; the female catkins mature quickly, releasing their seeds in late spring.

GERMINATION AND REGENERATION: The wind- and water-dispersed seeds germinate immediately on bare ground that is exposed after spring water levels recede; saplings and mature trees sprout readily from the base.

HABITAT PREFERENCES: River birch grows best in moist soil with full sun and prefers the edges of freshwater wetlands and pond, stream, and river banks. It is one of the most heat tolerant of all birch species, a trait that predisposes it to grow well in urban environments.

ECOLOGICAL FUNCTIONS: Heat reduction in paved areas; tolerant of roadway salt and compacted soil; stream and river bank stabilization.

CULTURAL SIGNIFICANCE: This species has become an important ornamental plant over the past 20 years, mainly because of its heat tolerance and its resistance to bronze birch borer and leaf miners, which often kill white-barked birches growing in urban areas. The cultivar 'Heritage' is widely planted because it maintains its attractive, buff-colored bark for a longer time than is typical for the species.

Mature river birch along the Charles River in Watertown, Massachusetts

River birch male (pendulous) and female (upright) flowers

River birch foliage

Distinctive bark of a river birch sapling

River birch stabilizing a riverbank

Betula populifolia Marsh.   Gray Birch

SYNONYMS: poverty birch, poplar birch, old field birch

LIFE FORM: **deciduous tree**; up to 35 feet (10 m) tall

PLACE OF ORIGIN: eastern North America

VEGETATIVE CHARACTERISTICS: Gray birch typically develops a multistemmed growth habit; its dull, chalky white bark is fairly tight (i.e., not peeling) and is marked with prominent dark, triangular patches below the branches. The alternate leaves are 2–3 inches (5–8 cm) long and more or less triangular (deltoid) with coarsely toothed margins; the leaves are glossy, dark green in summer and turn yellow in the fall. Individual gray birch stems are short-lived; they typically die when they reach a diameter of about 6 inches (15 cm).

FLOWERS AND FRUIT: Male catkins are produced at the ends of the branches and shed their wind-dispersed pollen in early spring; the female catkins produce cylindrical "cones" of seeds, 0.75–1.25 inches (2–3 cm) long, which disintegrate over the course of the fall and winter.

GERMINATION AND REGENERATION: The wind-dispersed seeds germinate in a variety of disturbed, sunny sites; mature trees typically sprout from the base to form a cluster of crooked or leaning stems.

HABITAT PREFERENCES: Gray birch is a pioneer species that is tolerant of poor soil and drought. It is mainly found growing on compacted or sandy, acid soil, often in association with quaking aspen. In the urban environment it is common on rock outcrops, stone walls, railroad rights-of-way, and the edges of rivers and streams.

ECOLOGICAL FUNCTIONS: Tolerant of roadway salt and compacted soil; erosion control on slopes; stream and river bank stabilization.

CULTURAL SIGNIFICANCE: As befits a tree known as the poverty birch, this species has very little cultural significance other than as an indicator or poor soil.

Mature, multistemmed growth form of gray birch

Male
(pendulous)
and female
(upright)
flowers of
gray birch

Gray birch foliage

Stand of gray birch along the Charles River in
Watertown, Massachusetts

Catalpa speciosa (Warder) Warder ex Engelm.   Northern Catalpa

Synonyms: Indian bean tree, cigar tree, western catalpa

Life Form: deciduous tree; up to 80 feet (25 m) tall

Place of Origin: central North America

Vegetative Characteristics: Northern catalpa has an upright growth habit with a strong central leader that is carried well up into a relatively narrow crown. It is coarsely branched with stout twigs, and the brownish bark has deep fissures separated by flaky ridges. The large, heart-shaped leaves taper to a sharp "drip-tip" and can be up to 14 inches (35 cm) long. The leaves are mostly opposite but can be whorled in groups of 3 toward the tip of the stem. The leaves are a distinctive light green during the growing season and turn pale yellow in the fall. The winter twigs display distinctive round leaf scars and prominent lenticels.

Flowers and Fruit: Northern catalpa produces upright, terminal panicles of showy white flowers in late May or June. The tubular blossoms, which are about 2 inches (5 cm) wide, have spreading lobes with wavy margins and are marked with 2 yellow stripes and numerous purple spots that serve as "nectar guides" for pollinating bees and moths. A mature, roadside tree in full bloom can be a spectacular sight. The slender fruits are more than a foot (33 cm) long but only half an inch (1 cm) wide; they are produced in abundance and persist on the tree through the winter, providing an unmistakable feature for identification. The pods contain numerous papery, thin seeds with a fringe of hairs at each end.

Germination and Regeneration: The wind-dispersed seeds germinate on bare soil in sunny locations; saplings sprout readily from the base following injury; the roots can produce shoots following traumatic injury.

Habitat Preferences: In its native habitat northern catalpa grows best in moist, sunny locations. In an urban context it occupies exposed sites including roadways, chain-link fences, foundation cracks, and river banks. Along highways it is a common colonizer of exposed rock cuts.

Ecological Functions: Heat reduction in paved areas; tolerant of roadway salt and compacted soil; erosion control on slopes; soil building on degraded land.

Cultural Significance: This midwestern native was widely planted in the East in the late 1800s because of its beautiful flowers, rapid growth, and rot-resistant wood that was used for fence posts. It fell out of fashion during the 1920s and 1930s when it developed a reputation for being "messy." Native Americans used a tea made from the bark as an antiseptic wash, a laxative, a sedative, and an antidote to snakebite; they made poultices from the leaves to treat wounds and bruises.

Related Species: The **common catalpa (*Catalpa bignonioides* Walt.)** is a smaller, less upright, and less hardy tree than northern catalpa; it is native to the Southeast and does not appear to naturalize in the Northeast.

Developing fruits of northern catalpa

A northern catalpa sapling growing on the grounds of an abandoned factory in Connecticut

Northern catalpa flowers

Northern catalpa foliage

Distinctive twig and leaf scars of northern catalpa

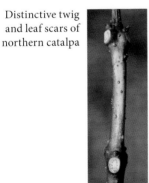

Fruits of northern catalpa remain on the tree through the winter

Lonicera japonica Thunb.   Japanese Honeysuckle

Synonyms: *Nintooa japonica*, Hall's honeysuckle

Life Form: **semi-evergreen vine**; up to 30 feet (9 m) long

Place of Origin: temperate eastern Asia

Vegetative Characteristics: Japanese honeysuckle climbs over other plants, smothering them with its luxuriant growth. The stems are reddish brown and hairy when young; older stems have flaky bark that peels off in strips. The ovate, hairy leaves are opposite and 1–3 inches (2.5–7.5 cm) long with entire margins. Plants are evergreen in the South and deciduous in the North.

Flowers and Fruit: Japanese honeysuckle produces tubular flowers in pairs on short stalks from May through September, depending on their latitude; the highly fragrant flowers are white at first but change to dull yellow as they age, producing a striking bicolor effect. The nectar-rich flowers are pollinated by bees and moths. The black fruits are about 0.25 inch (6 mm) wide, mature from late summer into fall, and are readily consumed and dispersed by birds.

Germination and Regeneration: Seeds germinate under a wide variety of conditions from sun to shade and moist to dry; runners and trailing stems root where they touch the ground, forming large clumps over time.

Habitat Preferences: Japanese honeysuckle is an early successional species that flourishes in disturbed open sites, floodplain thickets, and woodland edges. In urban sites it is found on chain-link fences, masonry walls, unmowed highway banks, and railroad rights-of-way. The plant grows most vigorously in full sun but is quite shade tolerant.

Ecological Functions: Tolerant of roadway salt; food and habitat for wildlife; erosion control on slopes.

Cultural Significance: Japanese honeysuckle was introduced from Europe in the early 1800s as an ornamental for its sweet fragrance and attractive flowers. The species has been spreading on its own since the early 1900s and is now a dominant species in disturbed woodland habitats throughout the Southeast and, more recently, the Northeast. Many states now list it as an invasive species. A tea made from the flowers is used for medicinal purposes in Asia.

Japanese honeysuckle in bloom in Boston

Japanese honeysuckle flowers

Japanese honeysuckle keeps its foliage in winter

The twining stems of Japanese honey suckle can overwhelm other plants

Japanese honeysuckle fruits

Japanese honeysuckle is evergreen in central New Jersey

Lonicera morrowii A. Gray   Morrow's Honeysuckle

SYNONYMS: none

LIFE FORM: deciduous shrub; 6–12 feet (1.8–3.6 m) tall

PLACE OF ORIGIN: Japan

VEGETATIVE CHARACTERISTICS: Morrow's honeysuckle is a multistemmed shrub with hollow stems and twigs and pale brown bark that peels off in thin strips. The opposite, dull green leaves are nearly hairless, oblong to egg shaped, and 1.5–2.5 inches (3–7 cm) long. Plants leaf out very early in the spring, well before most native shrubs do, and keep their leaves long after most native species have shed theirs in the fall. Morrow's honeysuckle thus has a growing season 2–4 weeks longer than that of native shrubs.

FLOWERS AND FRUITS: Morrow's honeysuckle flowers abundantly in May and June. The nectar-rich flowers, which are pollinated by a wide variety of insects, are initially white but become yellow as they age; they are followed in August by pairs of round red fruits about 0.25 inch (6 mm) wide that often rest on top of the leaves. Under full sun conditions a plant can be totally covered with fruit.

GERMINATION AND REGENERATION: The red berries are readily consumed and dispersed by birds, and seeds will germinate under a wide range of soil and light conditions. Stems root where they touch the ground, creating large, genetically uniform clumps.

HABITAT PREFERENCES: Morrow's honeysuckle grows best in full sun and moist soil but will tolerate drought and shade. It is common along disturbed woodland edges, wetlands, swamps, railroad tracks, rock outcrops, and roadsides.

ECOLOGICAL FUNCTIONS: Tolerant of roadway salt; food and habitat for wildlife; erosion control on slopes.

CULTURAL SIGNIFICANCE: Morrow's honeysuckle was introduced in 1875 as an ornamental. It was widely planted in the 1920s and 1930s for soil conservation purposes and escaped to become a common roadside plant. Many states now list it as an invasive species.

RELATED SPECIES: Tatarian honeysuckle (*Lonicera tatarica* L.), a native of central Asia and southern Russia that was introduced into North America in 1752, is somewhat similar in appearance to Morrow's honeysuckle, but produces pink rather than white flowers.

Morrow's honeysuckle in flower

Morrow's honeysuckle foliage

Flowers of Morrow's honeysuckle

Morrow's honeysuckle produces
abundant, bird-dispersed fruits

Morrow's
honeysuckle
fruits

Tatarian honeysuckle flowers
(photo by Les Mehrhoff)

Celastrus orbiculatus Thunb.   Oriental Bittersweet

Synonyms: Asiatic bittersweet, round-leaved bittersweet

Life Form: deciduous woody vine; stems up to 60 feet (18 m) long

Place of Origin: northeast Asia

Vegetative Characteristics: The twisted, woody stems are typically brown and have warty lenticles; with age they turn gray and develop into stout trunks up to 4 inches (10 cm) in diameter. Most of the alternate, simple leaves are 2–3 inches (5–7.5 cm) long by 1–2 inches (2.5–5 cm) wide, rounded to egg shaped with a short-pointed tip, but those produced by vigorous first-year stems can have long, tapering tips. Oriental bittersweet lacks tendrils and climbs by means of its twining stems, which eventually strangle its "host." The roots are bright orange. The leaves turn a clear, bright yellow in the fall.

Flowers and Fruit: This species produces separate male and female individuals (dioecious). Both sexes have small clusters of inconspicuous, insect-pollinated flowers in the leaf axils in May and June; female plants produce round green fruits that become highly conspicuous in the fall when they turn yellow and then split open to reveal orange-red seeds.

Germination and Regeneration: Bittersweet seeds are widely dispersed by birds and germinate readily under shady conditions. The stems can produce roots where they touch the ground, and roots can produce numerous shoots (suckers) following damage to the aerial stems. Once established, bittersweet is extremely difficult to eradicate because of its root-suckering ability.

Habitat Preferences: This highly adaptable vine can grow under a wide variety of light and soil conditions; young plants are extremely shade tolerant. Bittersweet is notorious for its capacity to strangle and overwhelm surrounding trees and shrubs, and it can be highly destructive in woodlands. In urban areas it is typically found on chain-link fences and stone walls, unmowed highway banks and median strips, along railroad tracks, and climbing up the trunks of low-branched trees and shrubs in minimally maintained landscape plantings.

Ecological Functions: Food and habitat for wildlife; erosion control on slopes.

Cultural Significance: Oriental bittersweet's attractive fruits—which are toxic—are widely used for holiday decorations, and the bark of the root is used in traditional Chinese medicine. The plant was introduced as an ornamental around 1860 and was planted to stabilize highway banks during the 1960s and 1970s. Since then it has spread rapidly throughout the East and is now listed as an invasive species in many states.

Strangling stems of Oriental
bittersweet

Oriental
bittersweet in fall
color in Boston

Oriental
bittersweet
foliage

Oriental bittersweet growing on chain-link fence

Mature fruits
of Oriental
bittersweet

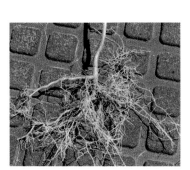

Bright orange roots of Oriental
bittersweet

Euonymus alata (Thunb.) Sieb.   Burning Bush

SYNONYMS: *Euonymus alatus*, winged euonymus, winged wahoo, winged staff tree

LIFE FORM: **deciduous shrub**; up to 16 feet (5 m) tall and equally broad

PLACE OF ORIGIN: temperate East Asia

VEGETATIVE CHARACTERISTICS: Burning bush has a wide-spreading, multi-stemmed growth habit; its green to brown branches have prominent corky wings that project out as much as half an inch (1 cm), creating a dramatic effect. The dark green leaves are opposite, elliptical, 1–3 inches (2.5–7.5 cm) long, and have very short petioles. In fall the leaves of plants growing in full sun turn brilliant red (hence the common name); those growing in shade turn rosy pink.

FLOWERS AND FRUIT: Small, greenish yellow, insect-pollinated flowers are produced from May to June. These are followed by reddish purple, 4-part fruit capsules, which split open in September to reveal 4 dangling orange-red "seeds" (the fleshy orange-red coat is technically an aril that surrounds the black true seed).

GERMINATION AND REGENERATION: Seeds are consumed and dispersed by birds; they germinate best in shady understory situations.

HABITAT PREFERENCES: Burning bush is a highly adaptable species that grows well in any combination of sun, shade, dry, or moist conditions as well as a broad range of soil pH. As a spontaneous plant in the urban environment, it is most common in the understory of emergent or disturbed woodlands.

ECOLOGICAL FUNCTIONS: Tolerant of road salt and compacted soil; food and habitat for wildlife; erosion control on slopes.

CULTURAL SIGNIFICANCE: Winged euonymus was introduced into North America as an ornamental plant around 1860. It is widely planted in a variety of low-maintenance landscape situations because of its spectacular fire-engine-red fall color. The cultivar 'Compacta', which typically stays below 6 feet (2 m) tall and has fewer "wings" on its branches, is especially common. Many states list the species as invasive.

SIMILAR SPECIES: Winged euonymus is much more commonly planted than European spindletree (*Euonymus europaea*), which is taller growing and altogether lacking fall color.

Growth
habit of a
spontaneous
burning bush
shrub

Burning bush flowers
and foliage

Burning bush mall planting in fall color

Spontaneous
burning bush
seedling under
a Norway
maple in fall
color

Burning bush fruits and fall foliage

Euonymus europaea L.   European Spindletree

SYNONYMS: European euonymus, skewerwood, prickwood

LIFE FORM: **shrub or small tree**; up to 25 feet (8 m) tall

PLACE OF ORIGIN: Europe

VEGETATIVE CHARACTERISTICS: European spindletree can grow as either an upright shrub or a small tree with an upright, rounded habit. It produces opposite branches and green twigs that lack corky wings. The shiny, opposite leaves are egg shaped to elliptic, taper to a sharp point, are 1–4 inches (2.5–10 cm) long by 0.5–1.5 inches (1.2–4 cm) wide and have a short petiole. The plant leafs out very early in the spring, and its fall color varies from yellow to reddish purple.

FLOWERS AND FRUIT: Small greenish or green-yellow, insect-pollinated flowers are produced in late April or May, followed in fall by pink, 4-part fruit capsules that open up to expose dangling orange-red "seeds" (the fleshy orange-red coat is actually an aril surrounding a black seed) that stay on the plant through November.

GERMINATION AND REGENERATION: Seeds are dispersed by birds; mature plants sprout vigorously from the base following injury.

HABITAT PREFERENCES: European spindletree grows well in moist, shady conditions, typically as an understory plant along the margins of disturbed woodlands; it can also be found at the base of stone walls and chain-link fences.

ECOLOGICAL FUNCTIONS: Tolerant of road salt and compacted soil; provides food and habitat for wildlife; erosion control on slopes.

CULTURAL SIGNIFICANCE: European spindletree was probably introduced into North America sometime during the 1700s and was reported to be "naturalized" in the mid-Atlantic region by 1811. It was widely planted as an ornamental into the twentieth century, but has mostly been displaced by burning bush (*Euonymus alata*) from Asia. The fruit has been used medicinally as a purgative.

SIMILAR SPECIES: European spindletree lacks the prominent corky wings and brilliant fall color that characterize winged euonymus (*Euonymus alata*); it also has a taller, more treelike growth habit.

WOODY DICOTS: **Celastraceae** (Stafftree Family)

European spindletree
holds its foliage well into
November in Boston

European spindletree flowers

European
spindletree
foliage showing
opposite leaf
arrangement

European spindletree has minimal fall color

European spindletree fruits

Elaeagnus umbellata Thunb.   Autumn Olive

SYNONYMS: Japanese oleaster, cardinal olive, Japanese silverberry

LIFE FORM: deciduous shrub; to 20 feet (6 m) tall

PLACE OF ORIGIN: temperate East Asia

VEGETATIVE CHARACTERISTICS: Autumn olive typically develops multiple gray trunks with smooth, gray bark; the smaller branches have stout thorns. The dull green leaves are alternate; 2–4 inches (5–10 cm) long; elliptical to egg shaped; and have smooth, wavy margins. The stems, the buds, and the undersides of the leaves are densely covered with silvery white to rusty brown scales. In late fall, the leaves turn dull yellow.

FLOWERS AND FRUIT: In May and June autumn olive produces clusters of fragrant, funnel-shaped, cream-colored to yellow flowers that are about 0.3 inch (1 cm) long; they are pollinated by bees and other insects. The flowers are followed by small, round fruits that are silvery at first with tiny brown scales, and turn red at maturity. The fruits are produced in great abundance, often completely covering stems.

GERMINATION AND REGENERATION: A wide variety of birds and other animals consume and disperse the fruits. Seeds germinate on exposed soil in unmowed grassy areas; established plants spout vigorously from the base following damage.

HABITAT PREFERENCES: Autumn olive is quite drought tolerant and grows best in full sun and sandy soils. At the present time it is more common along roadsides in the suburbs and the countryside than in urban areas, but this could change in the future. It is common along roadsides, forest edges, and abandoned pastures and lawns. A symbiotic relationship with filamentous bacteria in the genus *Frankia* results in the development of root nodules that "fix" atmospheric nitrogen into ammonia, allowing the plant to grow in nutrient-poor soils.

ENVIRONMENTAL FUNCTIONS: Food and habitat for a wide variety of birds and mammals; erosion control on slopes; soil improvement on degraded land.

CULTURAL SIGNIFICANCE: Autumn olive was widely planted along interstate highways for erosion control, wildlife habitat, and ornamental purposes from the 1960s through the 1970s. It has been spread widely by birds, and many northeastern states now list it as an invasive species. The edible fruits can be used to make an interesting jelly.

Autumn olive taking over in an abandoned baseball field in Vernon, Connecticut

Autumn olive flowers

Autumn olive flowers and foliage

Autumn olive growth habit

Autumn olive fruits are
attractive to birds

Amorpha fruticosa L.   Leadwort

SYNONYMS: false indigo, indigo bush, bastard indigo

LIFE FORM: deciduous shrub; up to 12 feet (3.6 m) tall

PLACE OF ORIGIN: North America

VEGETATIVE CHARACTERISTICS: Leadwort is a tall, multistemmed shrub with alternate, compound leaves consisting of 5–15 pairs of gray-green, oval leaflets, each about 0.75–1.25 inches (2–4 cm) long. The leaves turn pale yellow before they fall.

FLOWERS AND FRUIT: Spikes (racemes) of bee-pollinated flowers about 3–8 inches (7–20 cm) long are produced in early summer. The individual flowers are quite small and are maroon with prominent white anthers; they are followed by clusters of small pods, each less than 0.5 inch (1 cm) long.

GERMINATION AND REGENERATION: Leadwort pods are dispersed by water. Seeds germinate on moist, bare soil; established plants sprout readily from the base to produce dense clusters of flexible stems.

HABITAT PREFERENCES: Leadwort grows best in full sun, especially along the margins of freshwater marshes, swamps, streams, and rivers. "Nitrogen-fixing" root nodules containing symbiotic *Rhizobium*-type bacteria allow leadwort to produce its own fertilizer on low-nutrient sites.

ECOLOGICAL FUNCTIONS: Stream and river bank stabilization, especially where views need to be preserved (it can be cut down to the ground with impunity); tolerant of salt and soil compaction; soil improvement.

CULTURAL SIGNIFICANCE: Leadwort's flexible stems have traditionally been used to make baskets and other woven objects. It has been widely planted for soil conservation purposes and has escaped from cultivation in the Northeast as well as in Asia and Europe.

Leadwort growing along a reinforced shoreline of the Hudson River in New York

Leadwort in flower

Close-up of leadwort flowers

Developing seedpods of leadwort

Leadwort's muted fall color

Gleditsia triacanthos L.   Honey Locust

SYNONYMS: sweet locust, thorny locust, honeyshucks

LIFE FORM: deciduous tree; 70–90 feet (21–27.5 m) tall

PLACE OF ORIGIN: central North America

VEGETATIVE CHARACTERISTICS: Honey locust can be readily identified by its dark green, once- or twice-compound leaves, which can be up to a foot (30 cm) long; the foliage is finely textured with individual leaflets varying between 0.75 and 1.5 inches (2–4 cm) in length. The trunk and branches of young wild trees are often armored with sharp, heavy thorns that can be up to several inches (5–10 cm) long. Mature trees typically develop a graceful, upright form with a flat crown. The leaves turn clear yellow in fall.

FLOWERS AND FRUIT: Honey locust is a dioecious species; both male and female individuals produce relatively inconspicuous, insect-pollinated flowers in late spring. Females typically produce heavy crops of twisted, flat green pods that turn deep reddish brown at maturity. The pods, which range in size from 6 to 18 inches (15–45 cm) long by 1 inch (2.5 cm) wide, typically remain on the tree through the winter. They contain numerous hard brown seeds embedded in a sticky, greenish pulp that tastes slightly sweet and is the origin of the plant's common name.

GERMINATION AND REGENERATION: The hard seeds require some form of scarification (e.g., passing through the digestive system of animals that feed on the pods) to germinate; established plants produce root suckers.

HABITAT PREFERENCES: Honey locust has a reputation as a tough, drought-resistant street tree, but in the urban environment spontaneous plants are usually found in sunny bottomland sites or on recently disturbed soil. Unlike its cousin the black locust (*Robinia pseudoacacia*), honey locust does not fix atmospheric nitrogen and does not grow as well in nutrient-poor soils.

ENVIRONMENTAL FUNCTIONS: Heat reduction in paved areas; tolerant of roadway salt and compacted soil; erosion control on slopes; food and habitat for wildlife.

CULTURAL SIGNIFICANCE: Most of the honey locust trees being grown as street trees today are thornless male selections (var. *inermis*) that are less messy and less dangerous than thorny, seed-grown trees—in essence, the street tree version has been neutered and declawed. In the countryside, farmers often leave wild trees standing so that cattle and hogs can feed on the nutritious pods.

RELATED SPECIES: **Silktree** or **mimosa** (*Albizia julibrissin* **Durz.**) is a medium-sized, deciduous tree up to 30 feet (10 m) tall that is native to parts of central Asia (Iran) and eastern China. It produces alternate, twice-compound leaves—up to 20 inches (50 cm) long with numerous tiny leaflets—and white or pink "powder-puff" flowers that bloom throughout the summer. The blossoms are followed by 4–6 inch (10–15 cm) long pods that are initially green but turn brown in fall. Silktree is surprisingly common in urban vacant lots and the edges of degraded woodlands as far north as New York City. It grows best in full sun and, because of its nitrogen-fixing ability, tolerates poor soil. Silktree germinates readily from scarified seed, and established trees produce root suckers, especially following winter damage or disease.

Honey locust grown from seed with stout spines on its trunk

Spontaneous female honey locust along the roadside

Variation in honey locust leaves: *left*, twice compound; *center*, once compound; *right*, both once and twice compound

Honey locust foliage and seedpods

Silktree in full flower

Silktree flowers, seedpod, and foliage

Robinia pseudoacacia L.   Black Locust

SYNONYMS: common locust, yellow locust, white locust, false acacia, treenail

LIFE FORM: deciduous tree; 30–70 feet (9–21 m) tall

PLACE OF ORIGIN: eastern North America

VEGETATIVE CHARACTERISTICS: Old black locusts have gray to brown, deeply furrowed bark; the young stems have pairs of sharp spines at the nodes. The alternate, pinnately compound leaves are 8–14 inches (20–35 cm) long, with 6–20 rounded, pale blue-green leaflets that either don't change color or turn light yellow before dropping in the fall. The plant is late to leaf out in the spring.

FLOWERS AND FRUIT: Black locust produces 6–12 inch (15–25 cm) long chains (racemes) of showy, extremely fragrant flowers in mid-to-late spring; the pealike white flowers are about an inch (2.5 cm) wide with a bright yellow center and are mainly pollinated by bees. The flat brown pods are 2–4 inches (5–10 cm) long and stay on the plant well into winter.

GERMINATION AND REGENERATION: Black locust seeds will germinate in sunny, dry sites, but most young sprouts are root suckers produced by established plants. Over time, a single individual can give rise to a thicket of stems that, if unchecked, will develop into a genetically uniform stand of trees.

HABITAT PREFERENCES: Black locust tolerates a wide range of soils—from saline roadsides and acid mine tailings to rich, loamy limestone—but its greatest competitive advantage is on dry and sandy sites in full sun. In the urban environment it is common in minimally maintained public parks, vacant lots, waste dumps, emergent woodlands, unmowed highway banks, and railroad rights-of-way. "Nitrogen-fixing" root nodules containing symbiotic *Rhizobium* bacteria allow black locust to generate its own fertilizer on low-nutrient sites.

ECOLOGICAL FUNCTIONS: Heat reduction in paved areas; tolerant of roadway salt and compacted soil; erosion control on slopes; soil improvement; food and habitat for wildlife.

CULTURAL SIGNIFICANCE: Black locust has been widely cultivated since the 1700s as an ornamental and a source of rot-resistant wood for fence posts and poles. It was one of the first American trees to be widely planted on degraded soils in Europe during the late 1700s and has naturalized throughout much of that continent. More recently it has been planted throughout temperate Asia (most notably in Korea), where it has also escaped from cultivation. Black locust was used as a street tree in New York City as early as 1751. In the species' native Appalachia, black locust timbers were used as supports in coal mines because of their great tensile strength and rot resistance. Native Americans used the root bark to induce vomiting, but the leaves, seeds, and bark are toxic to both humans and livestock. The flowers are reportedly eaten in both Europe and Asia. Numerous horticultural varieties and hybrids of *Robinia* have unusual growth habits or colorful foliage. Black locust is widely planted on waste reclamation sites, including landfills and mine tailings.

Stand of mature black locust in Riverside Park, New York City

Black locust in bloom in late spring

Close-up of black locust flowers

Black locust foliage

Stout thorns on black locust stems

Stand of black locust stems developed from root sprouts

Quercus palustris Muenchh.   Pin Oak

SYNONYMS: none

LIFE FORM: deciduous tree; 60–70 feet (18–21 m) tall

PLACE OF ORIGIN: eastern North America

VEGETATIVE CHARACTERISTICS: Young pin oaks develop a pyramidal growth habit with a prominent central leader and horizontal side branches; on older trees the dead lower branches often persist and form a pendulous "skirt" around the trunk. The shiny, green, alternate leaves are 3–6 inches (8–15 cm) long by 2–4 inches (5–10 cm) wide, and have 5–7 sharp-pointed lobes that are separated by deep, U-shaped sinuses or indentations; their fall color varies from brown to bronze to deep red. The bark is grayish brown and smooth; with age it develops shallow ridges and furrows. Young trees typically hold onto their brown leaves through the winter, an important aid to identification. Pin oak has a shallow, fibrous root system that makes it easier to transplant than other oaks.

FLOWERS AND FRUITS: Delicate racemes of light yellow, wind-pollinated flowers are produced in the axils of the leaves in mid-spring. The relatively small acorns are about 0.5 inch (1–1.5 cm) long and mature at the end of their second growing season.

GERMINATION AND REGENERATION: Squirrels and birds consume and disperse the acorns when they ripen in October. Acorns that escape predation germinate the following spring. Saplings resprout vigorously from the base following injury or displacement from the vertical orientation.

HABITAT PREFERENCES: As a spontaneous tree in the urban environment (usually escaped from cultivation) pin oak is common along the edges of minimally maintained open space especially the margins freshwater wetlands, ponds, and streams; and along chain-link fences. It grows best with acid soil conditions and can tolerate sites that are flooded in winter (when it is dormant) and dry during the summer.

ECOLOGICAL FUNCTIONS: Heat reduction in paved areas; tolerant of roadway salt and compacted soil; food and habitat for wildlife; erosion control on slopes; stream and river bank stabilization.

CULTURAL SIGNIFICANCE: Pin oak is an important street tree in the East and, despite its preference for bottomland habitats is quite drought-tolerant. It also survives transplanting better than most other oaks. As a cultivated tree, this species has few peers in terms of adaptability, durability, and attractiveness.

SIMILAR SPECIES: Red oak (*Quercus rubra*) does not have the strong central leader carried up to the top of the crown, and the U-shaped indentions on its leaves are not nearly as deep as those of pin oak.

Pin oak growth habit

Pin oak foliage

Fall foliage of a mature pin oak in a Boston wetland

Immature pin oaks hold onto their withered leaves throughout the winter

Pin oaks in late autumn

Pin oak acorns

Quercus rubra L.   Red Oak

SYNONYM: northern red oak

LIFE FORM: deciduous tree; 60–90 feet (18–28 m) tall

PLACE OF ORIGIN: eastern North America

VEGETATIVE CHARACTERISTICS: The bark of a mature red oak consists of rough brown or black ridges separated by smooth, light patches of truck. The alternate, dark green leaves have a glossy appearance, are about 6 inches (15 cm) long by 4 inches (10 cm) wide, and have 7–11 bristle-tipped lobes separated by U-shaped indentations that extend about halfway to the main vein. Red oaks typically develop a single thick trunk and a rounded crown with wide-spreading branches. The foliage is late to color in the fall and turns anywhere from deep red to yellow-brown depending on the weather.

FLOWERS AND FRUIT: Red oak produces delicate, dangling chains—up to 5 inches (13 cm) long—of light yellow, wind-pollinated flowers in mid-spring, when the leaves are half grown and still pink. In the fall the plant produces 1 inch (2.5 cm) long, medium brown acorns that take 2 years to mature; they have a shallow, saucer-like cap that encloses only the base of the nut.

GERMINATION AND REGENERATION: Heavy crops of acorns (known as *mast*) are produced every 3–5 years; they are eagerly consumed by birds and mammals and are frequently cached by squirrels as a winter food supply. Uneaten seeds germinate in spring. Saplings growing in the forest understory resprout readily following injury, producing 3–4 inch (7–10 cm) tall "seedlings" that can be 20 or 30 years old. Large trees can also resprout following damage to their trunk.

HABITAT PREFERENCES: Red oak is a relatively fast-growing upland species that is widely cultivated along streets and in parks. It grows best in acid soils and full sun but is tolerant of air pollution, drought, and shade. It reproduces spontaneously in disturbed urban woodlands and along highway and railroad banks.

ECOLOGICAL FUNCTIONS: Heat reduction in paved areas; food and habitat for wildlife; erosion control on slopes; stream and river bank stabilization.

CULTURAL SIGNIFICANCE: Red oak is widely planted as a shade tree in urban environments; its lumber is very valuable, and its tannin-rich foliage deters feeding by insects. Native Americans used the acorns to make bread, but only after boiling them to remove the bitter taste.

SIMILAR SPECIES: Pin oak (*Quercus palustris*) typically produces a strong central leader that is carried well up to into the crown, and the U-shaped indentions on its leaves are much deeper than those of red oak.

Spontaneous red oak "street tree" in Boston

Red oak flowers and expanding spring foliage

Red oak bark

Red oak acorns

Red oak foliage

Morus alba L.   White Mulberry

SYNONYM: Common mulberry

LIFE FORM: Deciduous tree; 30–60 feet (9–18 m) tall

PLACE OF ORIGIN: eastern Asia

VEGETATIVE CHARACTERISTICS: The bark of young mulberry stems is light gray in color, often with a slight orange or yellowish tint; bark on older trees is brown and furrowed. The simple leaves are alternate, glossy, and 2–6 inches (5–15 cm) long; they have coarsely toothed margins and are highly variable in shape, ranging from unlobed to having up to 5 or 6 lobes separated by deep sinuses; indeed, this heterogeneity in leaf shape is a key identification feature. The young twigs exude milky sap when broken, and the leaves turn a clear yellow in the fall.

FLOWERS AND FRUIT: White mulberry produces inconspicuous greenish yellow, wind-pollinated flower spikes (catkins) in early spring. Male and female flowers are produced separately on the same tree (monoecious). The fruits are very juicy and turn black or purple (occasionally pink or white) at maturity. Roughly the size and shape of a blackberry, they ripen in mid-to late-summer; they are edible but rather insipid.

GERMINATION AND REGENERATION: White mulberry fruits are readily consumed and dispersed by birds, and seedlings readily sprout up beneath roosting sites. Damaged trees resprout vigorously from the base of the trunk.

HABITAT PREFERENCES: White mulberry is a highly adaptable, drought-tolerant species that grows best in full sun. In the urban environment it is common in minimally maintained public parks, disturbed or emergent woodlands, chain-link fence lines, stream and river banks, unmowed highway banks and median strips, and sidewalk and foundation cracks.

ECOLOGICAL FUNCTIONS: Heat reduction in paved areas; tolerant of roadway salt and compacted soil; food and habitat for wildlife; erosion control on slopes.

CULTURAL SIGNIFICANCE: White mulberry has been cultivated for several thousand years in China for its leaves, which are fed to silkworm caterpillars. In addition, the fruit is eaten raw and a medicinal tea is made from the leaves after they fall from the tree. White mulberry was introduced into North America in the mid-1700s with the intent of establishing a silkworm industry that never materialized. In the 1820s the silkworm craze reemerged when the dwarf white mulberry variety *multicaulis* was introduced. In *American Weeds and Useful Plants* (1959), William Darlington reported "a sort of *Multicaulis monomania* (or *Moro-mania!*)—so universal, and engrossing, that it became absolutely ludicrous; and was scarcely exceeded in absurdidty by the nearly contemporaneous epidemic which afflicted the nation in reference to its financial concerns. Almost everybody was eagerly engaged in cultivating myriads of trees . . . without stopping to enquire where they could be sold, or who would be likely to buy!" With the help of birds, white mulberry has become naturalized in a wide variety of urban habitats throughout the Northeast.

A mature white mulberry on a Detroit street

Typical white mulberry growth form

White mulberry leaves are highly variable in shape

White mulberry seedlings under stress often produce small leaves

White mulberry fruits at various stages of maturity

Frangula alnus P. Mill. Glossy Buckthorn

Synonyms: *Rhamnus frangula*, alder buckthorn, black dogwood

Life Form: deciduous shrub or small tree; 6–16 feet (2–5 m) tall

Place of Origin: Europe, North Africa, and central Asia

Vegetative Characteristics: Glossy buckthorn has smooth gray bark that is covered with prominent white or gray lenticels; it typically develops a multi-stemmed growth habit. The alternate, glossy green leaves are 1–2.5 inches (2.5–6 cm) long, round to oval in shape, and have prominent parallel veins and entire margins. The leaves maintain their dark green color well into fall, long after native species have gone dormant. Fall color, when it finally happens, is a dull yellow. The roots have a distinctive maroon color.

Flowers and Fruit: Glossy buckthorn produces small, bisexual, 5-petaled flowers in leaf axils throughout the summer; they are pollinated by bees. The flowers develop into 0.25 inch (6–7 mm) round fruits that are initially green then turn red before ripening to black in August and September. The simultaneous presence of flowers and multicolored fruits is a key feature that helps to distinguish glossy buckthorn from common buckthorn (*Rhamnus cathartica*). Among woody plants, glossy buckthorn is highly unusual in ripening its fruits over a period of several months.

Germination and Regeneration: The fruits are readily consumed by birds, and the seeds germinate under a wide variety of conditions from sunny and dry to shady and moist. Secondary stems sprout spontaneously from the base.

Habitat Preferences: Glossy buckthorn grows best in humus-rich soils and is abundant in the understory of minimally maintained woodlands; along the margins of freshwater wetlands, streams, and rivers; in vacant lots and rubble dumps; along chain-link fences; and on roadway edges and railroad rights-of-way.

Ecological Functions: Nutrient absorption in wetlands; tolerant of roadway salt; food and habitat for wildlife; erosion control on slopes.

Cultural Significance: The stem bark has a long history of medicinal use in Europe both as a laxative and to induce vomiting. The species was introduced into North America sometime before 1800 primarily for medicinal purposes. Several states list glossy buckthorn as an invasive species.

Similar Species: Common buckthorn (*Rhamnus cathartica*) is a dioecious species that produces clusters of black fruits that ripen at the same time in the autumn; its flowers have 4 rather than 5 petals; and its alternate leaves have toothed margins.

Glossy buckthorn
foliage and fruits

Glossy buckthorn leafing out early in a woodland
understory

Glossy buckthorn flowers

Multistemmed growth form of glossy buckthorn

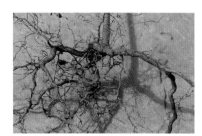

Red roots of glossy buckthorn

Glossy buckthorn fruits mature
continuously throughout the growing
season

Rhamnus cathartica L.   Common Buckthorn

Synonyms: hartshorn, waythorn, highwaythorn, ramsthorn, European buckthorn

Life Form: deciduous shrub or **small tree**; 10–25 feet (3–8 m) tall

Place of Origin: Eurasia and North Africa

Vegetative Characteristics: In cultivation common buckthorn may have a single trunk, but as a spontaneous plant it is usually multistemmed. The outer bark is gray, the inner bark is bright yellow, and the twigs are tipped with sharp thorns. The simple, glossy, dark-green leaves are mostly opposite with finely serrated edges; they are about 1.5–3 inches (4–8 cm) long and are elliptical to egg shaped. Plants leaf out early in spring and retain their green leaves much longer than the surrounding vegetation in autumn—falling only after they experience a hard frost.

Flowers and Fruit: Common buckthorn produces clusters of 4-petaled greenish yellow flowers in spring on separate male and female plants. Following pollination by insects, female plants produce clusters of small black fruits, about 0.25 inch (5 mm) wide, in September.

Germination and Regeneration: Fruits are dispersed by birds, and seeds will germinate under a wide variety of conditions; mature plants sprout readily from the base.

Habitat Preferences: Common buckthorn grows best in partially shaded woodland edges in soil with a neutral pH. It is also common along the margins of freshwater wetlands, ponds, and streams; in degraded woodlands; and on roadsides.

Ecological Functions: Tolerant of roadway salt and compacted soil; food and habitat for wildlife; erosion control on slopes.

Cultural Significance: Common buckthorn was included in Dioscorides' first-century herbal, *De Materia Medica*. Its bark and berries have long been used in traditional European medicine as a laxative; to induce vomiting; and to treat gout, dropsy, and rheumatism. Its sap has been used to make dyes. Common buckthorn was introduced into North America during the 1700s, probably for both medicinal and horticultural purposes (hedges), and by 1811 was reported to be "naturalized" in the mid-Atlantic region. It is seldom planted today, and many states—especially in the Midwest—list it as an invasive species.

Similar Species: The black fruits of common buckthorn mature at the same time in the fall, as opposed to the reddish or purple fruits of glossy buckthorn (*Frangula alnus*), which ripen continuously throughout the growing season. Also, glossy buckthorn has flowers with 5 petals, leaves with entire margins, and stems that lack thorns.

Growth habit of a mature common
buckthorn

Common buckthorn growing along a stream
in Boston in the fall

Common buckthorn flowers in spring

Common buckthorn foliage stays green
after the Norway maple and pin oak
surrounding it have turned

Common buckthorn
fruits ready for bird
dispersal

Malus species   Wild Apples

SYNONYMS: The identification of spontaneous apple trees is usually uncertain given the hybrid nature of most eating and ornamental apple varieties. Field manuals typically describe naturalized plants as belonging to the common apple group, *Malus pumila* P. Miller, but *M. baccata* (L.) Borkh. and *M. sieboldii* (Regal) Rehder have also been reported as naturalized species.

LIFE FORM: deciduous tree; 15–40 feet (5–12 m) tall

PLACE OF ORIGIN: Eurasia

VEGETATIVE CHARACTERISTICS: Wild apples typically grow as small, multi-stemmed trees or shrubs, usually broader than tall, with rough, scaly bark. The alternate, simple leaves are 2–4 inches (5–10 cm) long by 1–2 inches (2.5–5 cm) wide; they can be entire, serrated, lobed, or deeply incised with varying degrees of pubescence and typically turn yellow in the fall.

FLOWERS AND FRUIT: Wild apples produce showy white or pink, insect-pollinated flowers in early spring; they are produced in clusters on short shoots and have 5 petals that are positioned above the ovary. The small fruits are produced in the fall and range in diameter from 0.25 to 1 inch or so (0.5–3 cm); they can be red, yellow, or green at maturity.

GERMINATION AND REGENERATION: The fruits are eagerly consumed by birds and mammals, which may deposit the seeds some distance from the tree. Seeds germinate in a variety of conditions from sun to shade; saplings readily regenerate new shoots following damage. Some individuals produce root suckers and develop into thickets.

HABITAT PREFERENCES: As a cultivated plant wild apples grow best in loamy, well-drained, acid soils in full sun; they can survive in understory shade, but will not produce many flowers. As a spontaneous species in urban areas it is common in the understory of disturbed woodlands, on the edges of minimally maintained public parks, on rock outcrops and stone walls, and along unmowed highway banks and railroad tracks.

ECOLOGICAL FUNCTIONS: Heat reduction in paved areas; tolerant of roadway salt and compacted soil; food and habitat for wildlife; erosion control on slopes.

CULTURAL SIGNIFICANCE: Numerous species, hybrids, and cultivars of species in the genus *Malus* are widely cultivated for consumption (including cider production) and ornamental purposes. A few species of apple are native to central North America, but most edible apple varieties were introduced from Europe, some arriving with the first settlers. Flowering crabapples were introduced in the nineteenth and twentieth centuries, initially from Europe and later from Asia. Spontaneous apples are, in general, a mongrel race of trees.

Wild apple growing on the old High Line in New York City

A wild apple in full flower in Detroit

Growth habit of a wild apple seedling

Wild apple growth habit

Wild apple flowers

Wild apples

Prunus serotina Ehrh.   Black Cherry

Synonyms: *Cerasus serotina*, rum cherry, wild black cherry, bird cherry

Life Form: deciduous tree; 40–75 feet (12–23 m) tall

Place of Origin: eastern North America

Vegetative Characteristics: Black cherry has dark bark that is distinctively marked with horizontal white lines (lenticels); with age the bark breaks up into rectangular plates that make it quite rough. The alternate, lance-shaped leaves are dark green and glossy with finely serrated margins; they are typically 2–5 inches (5–12.5 cm) long by about an inch or so (2.5–4 cm) wide and turn yellow or pale pink in fall. The twigs and foliage smell like bitter almonds (prussic acid) when crushed. While black cherry is a tall, straight-growing tree in the forest, in urban environments it typically has a contorted, irregular form. The twigs and small branches are often infected with swollen, woody tumors caused by the black knot fungus.

Flowers and Fruit: In May and early June black cherry produces 4–5 inch (10–12.5 cm) long chains of insect-pollinated, 5-petaled, white flowers that terminate the leafy shoots. These are followed in August and September by dangling clusters of fleshy, purple-black cherries about 0.35 inch (8 mm) in diameter.

Germination and Regeneration: Birds eagerly consume black cherry fruits, and seeds germinate beneath their roosts in a variety of habitats from sun to shade and moist to dry. Saplings and established plants resprout vigorously from the base following injury.

Habitat Preferences: Black cherry grows best in full sun and fertile soil but tolerates drought and some shade. In the urban environment it is common along the margins of minimally maintained public parks, in the understory of disturbed or emergent woodlands, on riverbanks and unmowed highway banks, along chain-link fence lines, and at the base of trees and shrubs in ornamental planting beds.

Ecological Functions: Heat reduction in paved areas; food and habitat for wildlife; erosion control on slopes.

Cultural Significance: The fruits are used to make jelly and wine ("cherry bounce"), and the wood is highly prized for furniture and cabinet making. The powdered inner bark has been used in traditional medicine to treat coughs and bronchitis (Dr. Wistars's Balsam of Wild Cherry was famous in the early nineteenth century, and Smith Brothers' Wild Cherry cough drops were popular in the twentieth). Livestock that eat wilted black cherry foliage can get sick or even die from the cyanide the leaves contain. The species was widely planted in the Netherlands and portions of central Europe from the early 1900s through the 1950s and is now considered an invasive species in several countries.

Related Species: Choke cherry (***Prunus virginiana*** L.) is a shrubby, multistemmed species that typically grows to about 15 feet (5 m) tall and has dull green leaves that are shorter and broader than those of black cherry. It flowers a bit earlier than black cherry, and its fruits—which are red at maturity rather than black—are very tart (hence the common name). The fall color is reddish purple.

Black cherry
growth habit

Black cherry
bark; note the
distinctive
horizontal
striations

Black cherry flowers

Black cherry foliage

Mature black cherry fruits

Choke cherry fruits

Rosa multiflora Thunb. ex Murr.   Multiflora Rose

SYNONYMS: Japanese rose, living fence, rambler rose

LIFE FORM: **climbing shrub**; stems 10–16 feet (3–5 m) long

PLACE OF ORIGIN: temperate East Asia

VEGETATIVE CHARACTERISTICS: The stems of multiflora rose are smooth, green or reddish, and bear scattered recurved thorns that readily catch hold of clothing or skin. The alternate, pinnately compound leaves have 5–11 leaflets that are 0.5–1.5 inches (2–4 cm) long with serrated edges and fringed stipules at their base; they retain their lustrous bright green coloration until late in the fall, finally turning pale yellow. While it is not really a vine, the arching stems of multiflora rose show a strong tendency to grow up through other plants and smother them. Under the right growing conditions it forms impenetrable thickets.

FLOWERS AND FRUIT: In May and June, multiflora rose is covered with pyramidal clusters of fragrant, white to pale pink, insect-pollinated flowers that are 1–1.5 inches (2.5–4 cm) wide with 5 notched petals and a yellow center. The flowers are followed by loose bunches of small, round or egg-shaped, dull red rose hips that are 0.25 inch (6 mm) long; the rose hips mature in late summer and remain on the plant through most of the winter.

GERMINATION AND REGENERATION: Fruits are consumed (and dispersed) by birds, and the seeds can germinate under a wide variety of soil and light conditions. They also can remain viable in the soil seed bank for up to 20 years. Established plants sprout prolifically from the perennial woody base and from underground stems (rhizomes);the tips of the arching stems root where they touch the ground, allowing plants to spread rapidly across the landscape.

HABITAT PREFERENCES: Multiflora rose grows in a wide variety of ecological conditions and soil types ranging from moist, sunny sites to dry, shady ones. It is common in vacant lots, rubble dumps, the margins of woodlands, urban meadows, stream and river banks, and along chain-link fences and railroad tracks.

ECOLOGICAL FUNCTIONS: Tolerant of roadway salt and compacted soil; food and cover for wildlife; erosion control on slopes.

CULTURAL SIGNIFICANCE: Multiflora rose was originally introduced into North America in the early 1800s from Europe; by the late 1800s plants were being directly imported from Japan. During the 1930s–1950s, the plant was widely distributed by the U.S. Soil Conservation Service for erosion control under the name "living fence." With the help of birds, multiflora rose has spread far and wide across the landscape, and many states now list it as an invasive species. The thornless cultivar 'Inermis' is still used as an understock for grafted hybrid tea roses.

Multiflora rose growth and flowering habit

Multiflora rose foliage

Multiflora rose as a
spontaneous ornamental in
Watertown, Massachusetts

Multiflora rose
flowers

Multiflora
rose fruits

Rubus allegh"eniensis Porter   Common Blackberry

SYNONYM: bramble

LIFE FORM: **deciduous shrub**; 3–8 feet (1–2.5 m) tall

PLACE OF ORIGIN: eastern North America

VEGETATIVE CHARACTERISTICS: Common blackberry produces erect, green or dark red canes that are covered with stout, sharp spines that can inflict a painful prick on the unwary; individual canes typically live for 2 years before fruiting and dying. The palmately compound leaves have 5 dark green leaflets, red petioles, and typically turn dark burgundy red in fall. The terminal leaflet is always larger than the lateral leaflets and has the longest petiole.

FLOWERS AND FRUIT: Clusters of showy white, insect-pollinated flowers more than an 1 inch (2.5 cm) wide terminate 2-year-old canes in late May and June. These are followed by sweet, black berries in July. Blackberries have a solid core (receptacle) when picked, as opposed to raspberries, which are hollow because the receptacle remains attached to the plant.

GERMINATION AND REGENERATION: Numerous animals, including humans, relish blackberries, and the seeds germinate in a variety of habitats from full sun to partial shade. New shoots can arise either adjacent to the old canes from perennial crowns or some distance away on deep, underground stems.

HABITAT PREFERENCES: Common blackberry tolerates a wide variety of soil and light conditions but grows best in deep, moist soil and full sun. In the urban environment it is common in minimally maintained public parks; vacant lots and waste dumps; unmowed meadows and fields; woodland edges and openings; and dry, sunny slopes.

ECOLOGICAL FUNCTIONS: Food and habitat for wildlife; erosion control on slopes.

CULTURAL SIGNIFICANCE: The fruit, which is usually produced in abundance, is prized for making pies, jams, and jellies. A tea made from the twigs and leaves has been used to treat diarrhea.

A vigorous stand of young blackberry shoots

Blackberry foliage

Blackberry bushes in full flower

Blackberry flowers

Blackberry fruits

Rubus flagellaris Willd.   Northern Dewberry

Synonym: whip dewberry

Life Form: prostrate, deciduous vine; about 1 foot (30 cm) tall

Place of Origin: eastern North America

Vegetative Characteristics: Dewberry is a completely prostrate vine with sharp, backward-curving prickles scattered along its creeping stems. The compound leaves typically have 3 leaflets (occasionally 5)—0.75–1.5 inches (2–5 cm) long—with toothed margins.

Flowers and Fruit: Like the blackberry, northern dewberry produces large white, insect-pollinated flowers—about 1 inch (2.5 cm) wide—in May and June; these are followed by reddish black fruit about 0.5 inch (1.5 cm) wide.

Germination and Regeneration: Dewberry fruits are widely dispersed by birds and small rodents, and the seeds germinate in a variety of disturbed habitats. New stems emerge from the perennial woody crown at the base of the plant, and trailing stems can form adventitious roots at their nodes.

Habitat Preferences: Dewberry is common in sunny, dry sites with sandy or compacted soil, including urban meadows and minimally maintained grasslands. People who walk through such habitats often become entangled in (and scratched by) the plants.

Environmental Functions: Food and habitat for wildlife; erosion control on slopes.

Related Species: Swamp dewberry (*Rubus hispidus* L.) is a shrubby species that grows to about 2 feet (70 cm) tall and forms large, dense colonies in sunny or shady, moist conditions. Its flowers are slightly smaller than those of northern dewberry, and its stems are bristly rather than prickly, making it much less unpleasant to encounter. Its dark green compound leaves typically have 3 leaflets. F. L. Olmsted planted swamp dewberry in many of the parks he designed in the late 1800s.

Prostrate growth habit of northern dewberry

Northern dewberry stems trailing over an embankment

Northern dewberry flowers

Northern dewberry fruits

Swamp dewberry growth habit in late fall

Rubus occidentalis L.   Black Raspberry

SYNONYMS: blackcap, thimbleberry

LIFE FORM: **small deciduous shrub**; 4–6 feet (1.5–2 m) tall

PLACE OF ORIGIN: eastern North America

VEGETATIVE CHARACTERISTICS: Black raspberry produces alternate, compound leaves, usually with 3 (sometimes 5) leaflets that are 2–4 inches (5–10 cm) long and pale green above and whitish below. During the growing season its stems are conspicuously covered with a chalky, bluish-white powder and are armed with scattered, hooked prickles; in winter the stems turn a distinctive purplish maroon.

FLOWERS AND FRUIT: Two-year-old black raspberry stems produce terminal clusters of small-petaled, white flowers, about 0.5 inch (1.5 cm) wide, in early summer. Following insect-pollination, delicious purple-black fruits develop in mid-to-late summer; the canes die after they produce fruit. The core (receptacle) remains behind when the berry is picked (the blackberry receptacle detaches with the fruit).

GERMINATION AND REGENERATION: Fruits are dispersed by birds and other animals, and seeds germinate in sun or shade; new stems arise from a perennial base; 2-year-old stems arch over and root at the tip, giving rise to new plantlets.

HABITAT PREFERENCES: Black raspberry is an adaptable species that can tolerate a wide variety of soil conditions in either sun or shade. In the urban environment it is common along the margins of fields or woodlands, in vacant lots and waste dumps, in the understory of open woodlands, and on dry slopes.

ECOLOGICAL FUNCTIONS: Food and habitat for wildlife; erosion control on slopes.

CULTURAL SIGNIFICANCE: The purple-black fruits are excellent for eating right off the bush and for making pies and jam. A tea made from the root has been used medicinally to treat diarrhea and stomach pain and as a "female tonic."

RELATED SPECIES: Wineberry (*Rubus phoenicolasius* **Maxim.**) is an Asian raspberry with arching stems that root at their tips, compound leaves with 3 leaflets, and small white flowers. The terminal leaflet, which is about 4 inches (10 cm) long, is larger than the lateral leaflets. The plant is easily recognizable because all of its parts are densely covered with reddish glandular hairs that glisten in the sunlight. It produces delicious red fruits in July. This species grows vigorously in moist, sunny sites, and several states list it as an invasive species.

Black raspberry stems are chalky white with scattered thorns

Black raspberry leaves typically have 3 leaflets

Black raspberry flower

Black raspberry fruits

Wineberry stems are covered with soft, red spines

Wineberry fruits

Phellodendron amurense Rupr.   Amur Corktree

Synonyms: *P. japonicum, P. sachalinense, P. lavallei*

Life Form: **deciduous tree**; up to 60 feet (18 m) tall

Place of Origin: northeast Asia

Vegetative Characteristics: Amur corktree is a fast-growing species with wide-spreading branches; a flat-topped crown; and corky, deeply ridged bark that is the source of its common name. The opposite, pinnately compound leaves are dark green, glossy, and smooth. They are roughly a foot (30 cm) long, have an odd number of leaflets (anywhere from 5 to 13), and turn bright yellow in early fall. The bright yellow roots are conspicuous on seedlings pulled out of the ground.

Flowers and Fruit: Amur corktree produces greenish yellow, insect-pollinated flowers in late spring on separate male and female individuals. These are followed by clusters of oily black fruits, some of which remain on the tree through the winter. Starlings and mourning doves feed on—and disperse—the seeds after they have fallen to the ground in early spring.

Germination and Regeneration: The seeds germinate readily on disturbed, open ground as well as in the shady forest understory. Damaged trees sprout readily from the base.

Habitat Preferences: Amur corktree tolerates both shade and drought as well as a variety of soil conditions ranging from acid to alkaline; it is becoming an increasingly common component of disturbed or emergent woodlands adjacent to areas where it has been planted. It can also be found on rock outcrops, at the base of stone walls, and in neglected ornamental planting beds. While not yet particularly common in urban areas, Amur corktree spreads rapidly into open areas adjacent to fruiting trees.

Ecological Functions: Heat reduction in paved areas; tolerant of roadway salt and compacted soil; food and habitat for wildlife; erosion control on slopes.

Cultural Significance: Amur corktree is native to northeastern Asia, most notably the watershed of the Amur River, which separates China from Russia; it was introduced into North America as an ornamental in the late 1800s. It has spread from established plantings with the help of birds, and several states now list it as an invasive species. The yellow roots are used in traditional Chinese medicine.

Mature Amur corktree with flat-topped growth habit

Amur corktree roots are yellow

Amur corktree foliage

Amur corktree bark

Amur corktree fall foliage and fruit

Amur corktree fruits can stay on female trees all winter

Populus deltoides Marshall   Eastern Cottonwood

SYNONYMS: eastern poplar, Carolina poplar, cotton tree, white wood

LIFE FORM: **deciduous tree**; 50–80 feet (15–25 m) tall

PLACE OF ORIGIN: eastern North America

VEGETATIVE CHARACTERISTICS: The bark is ash gray to brown and, on old trees, is corrugated with deep ridges and furrows. Eastern cottonwood is a fast-growing species, and young trees typically display a leggy, sparsely branched form with smooth, greenish yellow stems. The young twigs are reddish and often have prominent corky "wings." The simple, alternate leaf blades are roughly triangular (deltoid), 3–5 inches (7–12.5 cm) long, and have coarsely serrated margins. The flattened petioles, which are often red, catch the slightest breeze, causing the leaves to quiver. Leaves on vigorous young sprouts often develop a vertical rather than horizontal orientation and are typically much larger than those found on the branches of older trees. The fall color is pale yellow.

FLOWERS AND FRUIT: Like all poplars, eastern cottonwood is dioecious, with separate male and female individuals. The wind-pollinated flowers are produced in early spring, well before the leaves come out. The pendulous male catkins are bright red and several inches (up to 7 cm) long; the female flowers are small and inconspicuous. In late spring the female capsules release thousands of tiny cottony white seeds, which accumulate in large masses around the base of the tree and are the source of the common name.

GERMINATION AND REGENERATION: Cottonwood's wind-dispersed seeds germinate without a dormancy delay in late spring. Seedlings grow rapidly on bare soil in full sun. Open-grown trees typically have multiple trunks, but it does not root sucker and form large clones like quaking aspen.

HABITAT PREFERENCES: In native habitats eastern cottonwood is a moisture-loving species that grows along the margins of freshwater wetlands, ponds, streams, and rivers. In urban areas it grows in disturbed wetlands and drainage channels, in vacant lots and rubble dumps, and along railroad tracks with ballast substrate.

ECOLOGICAL FUNCTIONS: Heat reduction in paved areas; tolerant of roadway salt; stream and river bank stabilization; food and habitat for wildlife.

CULTURAL SIGNIFICANCE: The inner bark contains the aspirin-like compound salicin, and Native Americans used it as a tea to treat scurvy and as a "female tonic."

SIMILAR SPECIES: Quaking aspen (*Populus tremuloides*) is much smaller in stature than black cottonwood and typically grows in dense thickets; the shape of its leaf is more rounded than triangular.

Mature eastern cottonwood in Boston

Eastern cottonwood resprouting following injury

Eastern cottonwood leaf in autumn, about 3 inches (7 cm) wide

Eastern cottonwood bark

Eastern cottonwood male catkins about to shed their pollen in early spring

Freshly dispersed eastern cottonwood seeds line a bike path in late spring

Populus tremuloides Michaux   Quaking Aspen

SYNONYMS: trembling aspen, American aspen, popple, chatterbox tree

LIFE FORM: **deciduous tree**; up to 50 feet (15 m) tall

PLACE OF ORIGIN: northern, central, and eastern North America

VEGETATIVE CHARACTERISTICS: Quaking aspen is a fast-growing, upright tree. The bark of young trees is smooth and yellow-green with black markings; mature trees have dark brown, furrowed bark. The leathery leaf blades are round or egg shaped with coarse teeth and are 2–4 inches (5–10 cm) long and wide. The distinctly flattened petioles cause the leaves to "tremble" in the slightest breeze. The leaves turn bright yellow in the fall.

FLOWERS AND FRUIT: Male trees produce long catkins that shed their wind-dispersed pollen in early spring before the leaves come out; female trees shed their tiny "cottony" seeds in late spring.

GERMINATION AND REGENERATION: Seeds germinate shortly after dispersal, on bare soil in full sun. Established trees reproduce vigorously from root suckers following injury, eventually forming large clumps of stems. In the West, quaking aspen groves consisting of thousands of trunks, all with the same genetic make-up, cover hundreds of acres and are thousands of years old.

HABITAT PREFERENCES: Quaking aspen is one of the most widespread trees in North America; it grows best in sunny, dry sites where competition from other trees is reduced. In the urban environment it is common on abandoned rubble dumps, in vacant lots, in open woods and thickets, and along railroad tracks and unmowed highway banks. In rural areas it is a common colonizer of disturbed sandy sites.

ECOLOGICAL FUNCTIONS: Food and habitat for wildlife; erosion control on slopes; soil building on degraded land.

CULTURAL SIGNIFICANCE: Quaking aspen bark contains the aspirin-like compound salicin, an anti-inflammatory agent, analgesic, and fever reducer. Native Americans used tea made from the bark to treat excessive menstrual bleeding, stomach pain, venereal disease, urinary aliments, worms, colds, and fevers. Today the tree is an important source of pulpwood for book and magazine paper.

RELATED SPECIES: **White** or **silver poplar** (*Populus alba* L.) is native to Europe. It produces alternate, simple leaves with 3–5 coarse lobes and distinctive fuzzy white undersides; the leaves are 2–4 inches (5–10 cm) long and nearly as broad, and turn yellow in the fall. This species was widely planted in the mid-1800s through the early 1900s as a windbreak but is not planted today. It exists mainly as a spontaneous plant that reproduces from root suckers in neglected urban woodlands and along the sunny margins of rivers and streams. In *American Weeds and Useful Plants* (1859), William Darlington reported that "some of the grass-plots in the public squares of New York have been quite overrun by the side-spreading suckers of this tree; even in closely-paved streets they work their way up between the stones. It should be discarded altogether." The Greek physician Dioscorides included white poplar in his first-century herbal *De Materia Medica*.

Mature stand of quaking aspen

Quaking aspen in the ballast along a Detroit rail line

Quaking aspen foliage

Quaking aspen colonizing a post-industrial meadow

White poplar trunk and branch

White poplar foliage; note the hairy white underside

Salix species   Pussy Willows

Synonyms: The numerous species of shrubby willows, including *Salix discolor, S. bebbiana*, and *S. purpurea*, can be very difficult to tell apart. Even the great naturalist Henry David Thoreau confessed, "The more I study willows, the more I am confused."

Life Form: multistemmed shrubs; 10–20 feet (3–6 m) tall

Place of Origin: eastern North America

Vegetative Characteristics: Pussy willows typically have a multistemmed habit; the smooth green, red, or yellow stems turn brown with age. The hairless leaves are typically oval to narrowly lanceolate and often have whitish or silvery undersides. The single smooth bud scale identifies winter buds. Leaves typically turn a nondescript yellow in the fall.

Flowers and Fruit: Willows are dioecious and their flowers can either be pollinated by wind or by insects. The male flowers (catkins) are showier than those of the females and are responsible for the familiar early spring display. The fruit capsules on female trees mature in late spring, releasing tiny seeds topped with silky hairs that facilitate wind dispersal. Pollination is typically by insects.

Germination and Regeneration: Willow seeds germinate in moist, sunny sites immediately after being shed. Established plants produce new stems from the base, and leaning stems will produce roots where they come in contact with the soil.

Habitat Preferences: Shrubby willows grow best along the sunny margins of freshwater wetlands, ponds, streams, and rivers and in roadside drainage ditches.

Ecological Functions: Nutrient absorption in wetlands; tolerant of roadway salt; food and habitat for wildlife; stream and river bank stabilization.

Cultural Significance: The supple stems are used for baskets and fences in Europe. The bark contains salicylic acid, a precursor of commercial aspirin, and is still used in traditional medicine. Shrubby willows have been used to phytoremediate sites contaminated with heavy metals (especially cadmium) as well as for biomass production in short-rotation forestry.

Shrubby willow growing in a moist microhabitat on the second story of an abandoned Detroit factory

Shrubby willow growing in a sunny wetland

Typical foliage of a shrubby willow with sharp-pointed lateral buds

Male flowers of a shrubby willow

The mature fruits of *Salix atrocinera* in spring

Paulownia tomentosa (Thumb.) Sieb. & Zucc. ex Steudel
Princess Tree

SYNONYMS: *Paulownia imperialis,* empress tree, karri-tree, royal paulownia

LIFE FORM: deciduous tree, up to 60 feet (18 m) tall

PLACE OF ORIGIN: temperate East Asia

VEGETATIVE FEATURES: *Paulownia* is fast-growing, sparsely branched tree with stout, pithy twigs. On mature trees the opposite, heart-shaped leaves are 6–12 inches (12–30 cm) long and nearly as wide, and are covered with velvety hairs, especially on the undersides; on juvenile trees the leaves can be up to 2 feet (70 cm) long and have 2 small secondary lobes. The leaves fall while green or turn brown after experiencing a hard freeze in autumn.

FLOWERS AND FRUIT: Prominent clusters of fuzzy brown flower buds develop at the ends of the branches in the fall. The buds remain in a rudimentary state of development throughout the winter, then expand in April or May to produce spectacular 2 inch (5 cm) long tubular flowers. The flowers, which are pale violet with dramatic yellow stripes on the inside of the corolla, are pollinated mainly by bees. They are followed by pointed, pecan-shaped woody capsules about 1.25–2 inches (3–5 cm) long that split open to release hundreds of small, wind-dispersed seeds in the fall. The spent pods often remain on the tree for several years. A mature *Paulownia* tree can produce up to 20 million seeds per year.

GERMINATION AND REGENERATION: The seeds germinate in spring on bare ground. Mature plants typically produce root suckers, especially following damage to the primary trunk. These shoots can arise at a considerable distance from the trunk, and most people assume they are seedlings.

HABITAT PREFERENCES: Princess tree is a light-demanding, drought-tolerant plant that grows in a variety of disturbed urban habitats, including vacant lots, chain-link fence lines, pavement and masonry cracks, rock outcrops, and highway and railroad banks. This species is currently most abundant in the mid-Atlantic region but can be expected to move farther north as the climate becomes warmer.

ECOLOGICAL FUNCTIONS: Heat reduction in urban areas; tolerant of roadway salt and compacted soil; erosion control on slopes; soil building on degraded land.

CULTURAL SIGNIFICANCE: *Paulownia* was introduced into North America in 1844. Its spread throughout the East was supposedly facilitated when seeds used as packing material to protect imported Chinese porcelain were discarded. The species is often hyped in Sunday newspaper supplements as a "wonder tree" that grows 6 feet (2 m) a year. Its light, fine-grained wood is highly valued in Japan for making a variety of specialized items.

SIMILAR SPECIES: Hardy catalpa (*Catalpa speciosa*) has smooth leaves and long, cigar-shaped fruits.

Princess tree seedling colonizing an abandoned building in New London, Connecticut

Growth habit of a spontaneous princess tree in New London, Connecticut

Princess tree saplings growing in good conditions produce huge leaves

Princess tree flowers

Winter twig of princess tree showing spent fruits and immature flower buds ready to open in spring

Ailanthus altissima (P. Mill.) Swingle   Tree-of-heaven

Synonyms: *Ailanthus glandulosa*, copal tree, ghetto palm, stink tree, Chinese sumac

Life Form: deciduous tree; up to 70 feet (21 m) tall

Place of Origin: central China

Vegetative Characteristics: The smooth, gray trunk is sparsely branched. The leaves are alternate, pinnately compound, can reach 2 feet (60 cm) in length, and are composed of 11–25 sharp-pointed leaflets 2–6 inches (5–15 cm) long. The leaves, which have an unpleasant odor when crushed, appear late in spring; the leaflets drop with pale yellow or no fall color after the first hard frost, leaving the bare leaf-stalk (rachis) behind. The stout, sparsely branched twigs with their prominent, heart-shaped leaf scars present a stark appearance in winter. The wood is very light and brittle with a core of brown pith.

Flowers and Fruits: *Ailanthus* produces small, greenish yellow flowers in June on separate male and female trees (dioecious). The male flowers have a noticeably unpleasant smell that attracts numerous insects that carry the pollen to female trees. Seeds contain an embryo that is located in the center of a slightly twisted, 1-winged seed (samara), and clusters of these wind-dispersed seeds persist on the tree into winter, making identification easy.

Germination and Regeneration: The seeds germinate readily on bare ground in sun or shade. The roots of mature trees can produce numerous suckers, especially after the primary trunk has been damaged or cut down; the stumps of young trees also sprout prolifically. In China the term "good for nothing *Ailanthus* stump-sprout" refers to a spoiled or irresponsible child who fails to live up to parental expectations.

Habitat Preferences: Tree-of-heaven grows well in full sun on dry, rocky or sandy soil, but is also quite shade tolerant. The tree is famous for its ability to tolerate disturbance and forms dense thickets of stems along unmowed railroad and highway embankments. It seems to have a special affinity for the cracks that form where concrete and blacktop come together. The foliage is reported to contain allelopathic chemicals that suppress the growth of nearby plants.

Ecological Functions: Extremely tolerant of stressful urban conditions including heat buildup, drought, air pollution, and road salt; important for slope stabilization and soil building on degraded land.

Cultural Significance: *Ailanthus* bark and leaves have a variety of uses in traditional Chinese medicine, including the treatment of diarrhea, dysentery, and tapeworm. *Ailanthus* was introduced into North America as an ornamental in the early 1800s and within 50 years had become naturalized in both urban and rural areas across most of the United States. Most states list the species as invasive. In 1859, Darlington noted that "there is no tree so generally employed in the city of New York as a shade tree." By the early 1900s *Ailanthus* had become thoroughly naturalized throughout the city and was famously celebrated in Betty Smith's 1943 novel, *A Tree Grows in Brooklyn*. In 1996 an unidentified fungal wilt disease was reported to be killing entire stands of *Ailanthus* in New York City.

Mature growth habit of a male tree-of-heaven

Tree-of-heaven growing through a chain-link fence

Tree-of-heaven leafing out in mid-spring

Tree-of-heaven is tolerant of shade and abuse

Seeds of the red-fruited tree-of-heaven, variety *erythrocarpum*

Solanum dulcamara L.   Bittersweet Nightshade

Synonyms: bittersweet, bitter nightshade, blue nightshade, poison berry, climbing nightshade, blue bindweed, violet bloom

Life Form: trailing, semiwoody vine; up to 10 feet (3 m) long

Place of Origin: Eurasia and North Africa

Vegetative Characteristics: The flexible green or purple stems of bittersweet nightshade trail along the ground or twine around and through other plants. The alternate, dark green leaves are 2–5 inches (5–12.5 cm) long and have 2 distinct forms: one is compound with 3 leaflets, of which the terminal segment is always the largest, and the other is simple and ovate to oval. The foliage has an unpleasant odor when crushed and plants growing in exposed, sunny sites often take on a dramatic dark purplish hue. The ropy roots are white and typically grow just below the soil surface.

Flowers and Fruit: Bittersweet nightshade produces clusters of small, star-shaped flowers along the stem, each with 5 purple to violet, downward-pointing (reflexed) petals and a cone of yellow anthers in the center. The contrasting colors of the flowers are attractive to pollinating bees. The mature fruit is an oval red berry about 0.5 inch (8–12 cm) long containing numerous small white seeds. Flowers and fruit are produced throughout the summer.

Germination and Regeneration: The fruits are dispersed by birds, and seeds germinate underneath roosts. Prostrate stems root where they touch the ground, and the shallow root system will produce suckers following traumatic injury to the main stem.

Habitat Preferences: Bittersweet nightshade is a drought-tolerant species that grows best in full sun and rich soil. In the urban environment it is common in vacant lots and rubble dumps; along chain-link fences, rock outcrops, and stone walls; and growing up through shrubs and trees in ornamental landscapes. In its native habitat it grows along the banks of rivers, streams, and lakes.

Ecological Functions: Tolerant compacted soils; food and cover for wildlife.

Cultural Significance: As is true of many members of the Solanaceae, the foliage of bittersweet nightshade is toxic and somewhat narcotic, although the ripe red fruits are harmless. The plant has a long tradition of use in European folk medicine to treat warts and other skin problems as well as jaundice and rheumatism. Modern science has confirmed that extracts show some anticancer activity. Darlington, writing in 1847, noted that bittersweet nightshade was "extensively naturalized in fertile soils, and is often tolerated and even sometimes cultivated to train over walls and fences, as its flowers and fruit are showy."

Bittersweet nightshade flowers and foliage often have a purplish cast

Bittersweet nightshade growing on a chain-link fence in Boston

Bittersweet nightshade foliage

Bittersweet nightshade flowers

Bittersweet nightshade fruits

Celtis occidentalis L.   Hackberry

Synonyms: raisin tree, sugarberry, northern hackberry, nettle tree, beaver wood

Life Form: deciduous tree; up to 60 feet (15 m) tall

Place of Origin: eastern North America

Vegetative Characteristics: Hackberry is a small to medium-size tree with a short trunk and an irregular crown; its gray-brown bark is distinctively covered with corky warts or ridges. The leaves are alternate, simple, 2–4 inches (5–10 cm) long by 1.5–2 inches (3.5–5 cm) wide, and rough to the touch; they are finely serrated, taper to a point at the tip, and are lopsided (unequal) at the base. Dense clusters of twigs known as "witch's brooms" typically develop on the upper branches. The leaves turn light yellow in fall.

Flowers and Fruit: Hackberry produces inconspicuous, wind-pollinated flowers in spring as the leaves are unfolding. Solitary female flowers are produced on slender stalks in the axils of the leaves; they are followed by small, dark purple fruits about 0.25 inch (5–7 mm) in diameter, each containing a single large seed. Fruits ripen in the fall and can persist on the tree through the winter.

Germination and Regeneration: Hackberry fruits are consumed and dispersed by birds, and the seeds germinate on a variety of sites. Established trees sprout readily from the stump, and roots will produce shoots following traumatic injury to the main stem.

Habitat Preferences: Hackberry is a highly adaptable species that can grow in a wide variety of habitats and exposures. In its native habitat it prefers limestone outcrops and moist bottomlands with good light; in the urban environment hackberry grows in dry, heavy or gravelly soils with either a high or a low pH. While not particularly common, hackberry can be found growing along the margins of streams and rivers, roadsides, rock outcrops and stone walls, and railroad tracks. Once established it can persist indefinitely.

Ecological Functions: Tolerant of roadway salt and compacted soil; food and habitat for wildlife; erosion control on slopes; heat reduction in paved areas.

Cultural Significance: Hackberry fruits are edible in the fall and are said to taste like raisins or dates. While seldom planted in the East, hackberry is commonly cultivated in the Midwest because it tolerates heavy, high pH soils.

Hackberry foliage

Hackberry growing in Central Park, New York City

Hackberry can be identified in winter by the distinctive "witches' brooms" in its crown

Hackberry fruits

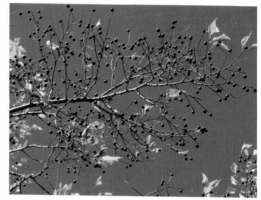

Hackberry's characteristic gray, warty bark

Ulmus americana L.   *American Elm*

SYNONYMS: white elm, water elm, soft elm

LIFE FORM: Deciduous tree; up to 100 feet (30 m) tall

PLACE OF ORIGIN: eastern North America

VEGETATIVE CHARACTERISTICS: American elm has a distinctive upwardly arching, vase-shaped form; older trees develop a broad, spreading crown. The bark is dark gray with broad, deep ridges. The leaves are alternate, oblong to egg shaped, and 3–6 inches (7–15 cm) long by 1–3 inches (2.5–7.5 cm) wide. They have doubly serrated margins, are rough to the touch, and have unequal bases, giving them a somewhat lopsided appearance. The leaves have short petioles and turn yellow in the fall.

FLOWERS AND FRUIT: American elm produces wind-pollinated, green to yellow flowers in very early spring. These are followed quickly by conspicuous clusters of papery "wafers" — about half an inch (13 mm) wide — that are dispersed by wind in early June.

GERMINATION AND REGENERATION: Seeds are shed in late spring and germinate immediately on moist ground; saplings sprout readily from the base following damage.

HABITAT PREFERENCES: As a wild plant American elm is a bottomland species that grows best in moist, sunny habitats. In the urban environment it grows on moist or dry compacted soils and is common in minimally maintained open space; vacant lots and waste dumps; emergent woodlands; along the margins of freshwater wetlands, ponds, streams, and rivers; chain-link fence lines; roadside drainage ditches; and railroad rights-of-way.

ECOLOGICAL FUNCTIONS: Heat reduction in paved areas; tolerant of roadway salt and compacted soil; erosion control on slopes; stream and river bank stabilization.

CULTURAL SIGNIFICANCE: This widespread species has a long history of cultivation in eastern North America because of its beautiful, vase-shaped form and because it is easy to transplant, and tolerates heavy pruning. Prior to the arrival of Dutch elm disease in the 1930s, American elm accounted for roughly 50% of the street trees in the Northeast. Mature specimens are now rare in the landscape, but young trees are common and typically survive long enough to produce seed. Several cultivars of American elm resistant to Dutch elm disease have been introduced into the nursery trade.

RELATED SPECIES: Siberian elm (*Ulmus pumila* L.) has a stiffer, more upright growth habit than American elm, and its leaves are much smaller, only about 1–3 inches (3–7 cm) long and about half as wide. It flowers in early spring and disperses its seeds soon after. Siberian elm, which is native to northeast Asia, was widely planted in the late 1800s and early 1900s as a windbreak and as a street tree in urban areas because of its drought and cold tolerance. It is seldom planted today because of its lack of ornamental features and perceived messiness (it sheds twigs and branches in the slightest wind). It grows on a variety of disturbed urban sites and can reach heights up to 80 feet (24 m).

Growth habit of mature American elm

Serrated leaf of American elm

Two American elms in Hartford, Connecticut, one pruned brutally, the other allowed to grow freely

Developing American elm fruits

American elm has alternate leaf arrangement

Siberian elm leafing out in mid-spring

Ampelopsis brevipedunculata (Maxim.) Trautv.   Porcelain Berry

SYNONYMS: *Ampelopsis heterophylla*, Amur peppervine, Asiatic creeper

LIFE FORM: deciduous woody vine; up to 20 feet (6 m) long

PLACE OF ORIGIN: northeast Asia

VEGETATIVE CHARACTERISTICS: The alternate, light green leaves may be up to 5 inches (12 cm) long and have hairy undersides. Leaves on the older parts of the stem (mature leaves) have 3 well-defined lobes; the juvenile leaves produced at the ends of the vigorously growing shoots are deeply dissected and irregularly toothed. The plant climbs by means of tendrils that are produced opposite the leaves; old plants have flexible stems with tight brown bark and prominent lenticels. The leaves typically drop without changing color.

FLOWERS AND FRUIT: Porcelain berry produces inconspicuous clusters of greenish white, 5-petaled flowers along its stems from July through August. They are insect-pollinated and are followed by hard, round fruits that are initially yellow, green, or lilac and eventually turn beautiful marbled shades of blue and white (the source of its common name) from September through October.

GERMINATION AND REGENERATION: Mature porcelain berry fruits are readily consumed by birds, which disperse the seeds across the landscape; trailing stems root where they touch the ground, facilitating the rapid spread of this vine.

HABITAT PREFERENCES: Porcelain berry grows in a wide variety of disturbed habitats in full sun to part shade, but it does best when it has access to good soil moisture. It is common along highway and railroad banks, forest edges and thickets, rock outcrops and stone walls, and chain-link fence lines—all habitats where it can overwhelm adjacent vegetation.

ECOLOGICAL FUNCTIONS: Food and habitat for wildlife; erosion control on slopes; stream and river bank stabilization.

CULTURAL SIGNIFICANCE: Porcelain berry was introduced from Asia in the mid-1800s as an ornamental plant for its beautiful blue fruits and finely dissected leaves. It escaped from cultivation and is now listed as an invasive plant in many states. The cultivar 'Elegans' has green and white variegated leaves.

Porcelain berry growth habit

Porcelain berry climbing a wall with its tendrils

Highly dissected juvenile leaves of porcelain berry

Mature leaves and fruit of porcelain berry

Ripening porcelain berry fruits

Parthenocissus quinquefolia (L.) Planchon   Virginia Creeper

SYNONYMS: woodbine, five-leaved ivy

LIFE FORM: **deciduous vine**; up to 60 feet (18 m) long

PLACE OF ORIGIN: eastern North America

VEGETATIVE FEATURES: The alternate, palmately compound leaves typically consist of 5 leaflets, 1.5–4 inches (4–10 cm) long, with coarsely toothed margins and a long petiole. The plant climbs by means of branched tendrils with oval, adhesive disks at their tips that produce a kind of "cement" that allows them to adhere tightly to a variety of surfaces. The leaves turn brilliant red in the fall.

FLOWERS AND FRUIT: Virginia creeper produces short spikes of inconspicuous greenish white, insect-pollinated flowers in July that are usually hidden by the leaves. These are followed by small blue-black berries in September and October.

GERMINATION AND REGENERATION: The fruits are dispersed by birds, and seeds germinate in sun or shade. Stems that contact the ground will form adventitious roots.

HABITAT PREFERENCES: Virginia creeper grows in a wide variety of urban habitats but is most often seen climbing trees along woodland edges; it is also common along dry, sunny dry roadsides and on chain-link fences.

ECOLOGICAL FUNCTIONS: Food and cover for wildlife.

CULTURAL SIGNIFICANCE: Native Americans used this plant for a variety of medicinal purposes, but mainly as an astringent and a diuretic. Virginia creeper has a long history of cultivation for its fall color in both North America and Europe.

RELATED SPECIES: **Boston ivy (*Parthenocissus tricuspidata* (Sieb. & Zucc.) Planchon)** is native to eastern Asia and produces 3-lobed, entire leaves—4–8 inches (10–20 cm) wide—that have a glossy sheen. It is often used to cover masonry walls and only occasionally spreads beyond the areas where it is cultivated when birds disperse the seeds. The foliage turns a dramatic deep red to maroon in the fall. The plant was first introduced into North America (Boston) from Japan in 1862. The term *Ivy League* supposedly comes from the fact that this plant covered the buildings of exclusive colleges in the Northeast.

Virginia creeper growing along a chain-link fence in Boston

Virginia creeper fruits

Virginia creeper's brilliant fall color

Virginia creeper leaves

Boston ivy growing under a dry, shady highway

Boston ivy foliage

Vitis riparia Michx.   Riverbank Grape

SYNONYMS: frost grape, wild grape

LIFE FORM: Deciduous woody vine; climbing up to 60 feet (18 m)

PLACE OF ORIGIN: eastern North America

VEGETATIVE CHARACTERISTICS: Riverbank grape produces simple, alternate leaves with 3 lobes and relatively few hairs on the underside; they are palmately veined with toothed margins and 4–8 inches (10–20 cm) long. The vines climb by means of forked, coiling tendrils that are produced along the stem opposite the leaves. With age, the dark brown stems can become several inches (5–15 cm) thick with bark shedding off in thin strips. Leaves turn yellow in the fall.

FLOWERS AND FRUIT: Separate male and female plants produce inconspicuous chains of greenish yellow, insect-pollinated flowers in late spring to early summer; they arise in the axils of the leaves and are about 4 inches (10 cm) long. The purple-black fruits are smaller than cultivated grapes, about 0.25–0.5 inch (6–12 mm) in diameter, and are produced abundantly by female plants in the fall.

GERMINATION AND REGENERATION: Riverbank grapes are eaten by a variety of birds, and the seeds germinate freely under their roosts; trailing stems root where they touch the ground.

HABITAT PREFERENCES: Seedlings are highly shade tolerant; once they reach the sunny forest canopy the vines spread out and begin flowering. Wild grapes are vigorous climbers that can easily overwhelm adjacent vegetation. In the urban environment they are common in the understory of moist woods and thickets, along the banks of streams and rivers, climbing chain-link fences, and on roadside guardrails. Like most vines, wild grapes grow best when their roots are situated in moist, shady soil and their leaves are in full sun.

ECOLOGICAL FUNCTIONS: Tolerant of roadway salt; food and habitat for wildlife; erosion control on slopes.

CULTURAL SIGNIFICANCE: The fruit is edible and makes excellent jelly. Native Americans made a tea made from the leaves for medicinal use. Established wild grapes can be very destructive to forest trees, particularly when weighted down by heavy, wet snow or ice.

RELATED SPECIES: The leaves of **fox grape** (*Vitis labrusca* L.) have a dense covering of brownish or whitish hairs on the underside, giving them a rusty or grayish appearance when they blow in the wind. The fruits are large, about 0.75 inch (2 cm) in diameter, and sweet. This species is one of the parents of the famous hybrid 'Concord' grape developed in 1852 by Ephraim Bull of Concord, Massachusetts.

Riverbank grape on a telephone line in Detroit

Riverbank grape overwhelming adjacent vegetation

Riverbank grape fruits

Tangle of riverbank grape stems

Riverbank grape foliage

The woolly undersides of fox grape leaves are distinctive

Mollugo verticillata L.   Carpetweed

SYNONYMS: Indian chickweed, devils-grip, whorled chickweed

LIFE FORM: **summer annual**; up to 18 inches (45 cm) wide

PLACE OF ORIGIN: Central America, possibly even North America

VEGETATIVE CHARACTERISTICS: The main branches of this prostrate, mat-forming plant radiate out like the spokes of a wheel and, through repeated branching, form a more or less circular mat. The narrow, light green leaves, which are widest at the tip and taper down to a narrow base, are 0.5–1.25 inches (1–3 cm) long; they are produced in whorls of 3–8 at each node, separated by discrete segments of smooth stem. Carpetweed produces a slender taproot and its branches do not root at the nodes.

FLOWERS AND FRUIT: Clusters of small, greenish white flowers with 5 petal-like sepals on slender pedicles arise in the axils of the whorled leaves from June through September. They can be either self- or insect-pollinated. The fruit is a small capsule containing numerous tiny brown seeds.

GERMINATION AND REGENERATION: Carpetweed seeds germinate readily in sunny sites, either moist or dry.

HABITAT PREFERENCES: In the urban environment carpetweed is abundant in sidewalk cracks and pavement openings, often growing in association with purslane and prostrate knotweed. It thrives in compacted or sandy soil in full sun and is common in ornamental planting beds, vegetable gardens, vacant lots, and rubble dumps.

ECOLOGICAL FUNCTION: Disturbance-adapted colonizer of bare ground.

CULTURAL SIGNIFICANCE: Young plants are purported to be edible, but one would have to be very hungry.

SIMILAR SPECIES: Spotted spurge, purslane, and prostrate knotweed are other mat-forming plants with a circular growth pattern and distinct taproot that seem "pre-adapted" to grow in sidewalk cracks and survive being stepped on.

Carpetweed growth habit

Carpetweed growing in sandy soil

Carpetweed flowers and foliage

Carpetweed surviving on air-conditioner drip in the urban environment

Carpetweed flowers

Amaranthus retroflexus L.   Redroot Pigweed

SYNONYMS: rough pigweed, green amaranth

LIFE FORM: summer annual; up to 6 feet (2 m) tall

PLACE OF ORIGIN: Central America

VEGETATIVE CHARACTERISTICS: The upright, hairy stems and stout taproot are typically suffused with red pigment. The leaves are alternate, egg or diamond shaped, have wavy margins, and can be up to 4 inches (10 cm) long. The uppermost leaves are typically lance shaped and smaller than the lower ones.

FLOWERS AND FRUIT: Redroot pigweed produces spikes of small, tightly clustered green flowers that form an erect terminal inflorescence. Separate male and female flowers are produced on the same plant and are mainly wind-pollinated. The flower heads develop into dense, bristly seed heads containing small, round black seeds enclosed in a brown sheath.

GERMINATION AND REGENERATION: Seeds germinate readily on bare sunny ground; in some agricultural fields there can be hundreds of seedlings per square yard (meter).

HABITAT PREFERENCES: Like most agricultural weeds, redroot pigweed prefers moist, rich soil and full sun. ECOLOGICAL FUNCTIONS: Disturbance-adapted colonizer of bare ground; winter food for wildlife (especially birds).

CULTURAL SIGNIFICANCE: Seeds and leaves of various amaranth species were an important source of food and medicine for Native Americans. Some people still gather the young leaves and shoots for spring greens. Prior to the Spanish conquest of South and Central American an amaranth species known as *kiwicha* (*Amaranthus caudatus*) was an important cultivated grain for the Aztecs and the Incas.

RELATED SPECIES: **Purple** or **livid pigweed** (*Amaranthus blitum* L.) is a low-growing annual or biennial of uncertain tropical origin. It forms a spreading mat with succulent green or red stems; the smooth leaves are up to 1.5 inches (4 cm) long with a prominent notch at their tip. Purple pigweed produces small greenish flowers from June through August in dense clusters at the ends of the branches and in the lower leaf axils. It is common in a variety of disturbed sunny sites and was once cultivated for its edible leaves.

Taproot of redroot pigweed

Redroot pigweed growth habit

Redroot
pigweed
(*left*) and
purple
pigweed
(*right*)

Redroot pigweed flowers

Purple pigweed produces leaves with a
notched tip

Purple pigweed growth habit

Daucus carota L.   Wild Carrot

SYNONYMS: Queen Anne's lace, bird's nest

LIFE FORM: herbaceous biennial; up to 3–4 feet (1–1.3 m) tall

PLACE OF ORIGIN: Eurasia and North Africa

VEGETATIVE CHARACTERISTICS: Wild carrot is a tall, slender plant with finely dissected, pinnately compound foliage that has an aromatic, carrot-like odor. During its first year the plant forms a rosette of bipinnately compound leaves—up to 6 inches (15 cm) long—that remain green through the winter; the second year it sends up a tall flowering stalk with alternate leaves. The stout, whitish taproot is difficult to pull out of the ground.

FLOWERS AND FRUIT: Wild carrot produces numerous lacelike white flowers in flat-topped, terminal clusters (umbels) from June through September; they can be insect- or self-pollinated. About one in four plants has a single deep purple flower (the "fairy seat") in the center of the cluster of all-white flowers. As the seeds develop, the umbels close up and develop a form resembling a bird's nest. The tiny seeds are covered with numerous barbs that facilitate their dispersal by animals. A single plant can produce up to 4,000 seeds, and the tall stalks are often bent over by their weight.

GERMINATION AND REGENERATION: The seeds germinate readily on disturbed, sunny sites in spring.

HABITAT PREFERENCES: Wild carrot tolerates full sun and dry soil. It is common in abandoned grasslands and urban meadows, vacant lots, rubble dumps, rock outcrops, stone walls, roadsides, and railroad rights-of-way. In its native European habitat it is common in coastal meadows.

ECOLOGICAL FUNCTIONS: Disturbance-adapted colonizer of bare ground; tolerant of roadway salt and compacted soil; food for wildlife.

CULTURAL SIGNIFICANCE: Seeds of wild carrot have long been used in European traditional medicine as a "morning-after" contraceptive, and in India to reduce female fertility. Indeed, Dioscorides' first-century herbal, *De Materia Medica*, clearly describes its anti-fertility properties. The use of wild carrot as a contraceptive has been documented in the Appalachian Mountains of North Carolina as well, passed down through oral tradition (Riddle, 1999). The plant has also been used as a diuretic to cure kidney and bladder stones and to eliminate worms. Although it is considered the ancestor of the domestic carrot, the roots are barely edible. In *American Weeds and Useful Plants* (1859), Darlington interpreted the presence of this plant as a sign of moral weakness: "When it gets on the premises of a careless, slovenly farmer, it soon multiplies so as to become a source of annoyance to the whole neighborhood."

Wild carrot growth habit

Wild carrot flowering on a roadside

Wild carrot foliage

Wild carrot flower head with "fairy seat" in the center

Developing seeds of wild carrot

Asclepias syriaca L. Common Milkweed

SYNONYMS: *Asclepias cornuti*, wild cotton, silkweed, cotton weed

LIFE FORM: herbaceous perennial; up to 3–4 feet (1–1.3 m) tall

PLACE OF ORIGIN: eastern North America

VEGETATIVE CHARACTERISTICS: Common milkweed produces stout, erect stems that are hollow and unbranched. The leaves are opposite, oblong to oval, and 3–6 inches (7.5–15 cm) long with a prominent white mid-vein; the upper surface of the leaf is smooth, the underside is covered with downy hair. All parts of the plant are covered with soft hairs and exude milky sap when broken.

FLOWERS AND FRUIT: Milkweed produces its fragrant, 5-petaled, pink-purple to white flowers from late June through early August. The distinctive round, drooping clusters attract a wide variety of insect pollinators. The fruits are 3–4 inch (7.5–10 cm) long, teardrop-shaped pods with prominently curved tips, soft spines, and a gravity-defying upward orientation. In October they split open to release hundreds of flattened black seeds, each topped with a tuft of soft, silky fibers to facilitate wind dispersal.

GERMINATION AND REGENERATION: Seeds germinate readily in open, sunny locations. Established plants produce new shoots from the base of the old stalks and from buds produced by the widely spreading roots, forming extensive colonies.

HABITAT PREFERENCES: Milkweed grows best in sunny, disturbed habitats without much regard for soil pH. It is common in infrequently mowed meadows, sandy roadsides, vacant lots, and along railroad tracks.

ECOLOGICAL FUNCTIONS: Host plant for larvae of the monarch butterfly; copious nectar production.

CULTURAL SIGNIFICANCE: Young milkweed shoots and leaves are edible when cooked. Native Americans used a tea made from the roots as a laxative and for treating kidney stones and dropsy, and made rope from its stem fibers and pillows from its silk. During World War II, the silky seed hairs were used as a substitute for kapok to fill "Mae West" life vests. Between 1943 and 1945, a million such flotation devices were filled with the floss from some 24 million pounds (11 million kg) of milkweed pods!

 Milkweed spreads
extensively from root
sprouts

Milkweed flowers

Milkweed foliage

Milkweed dispersing seeds

Milkweed fruits

Synonyms: *Cynanchum nigrum, Vintoxicum nigrum,* climbing milkweed, Louise's swallowwort

Life Form: herbaceous perennial vine; 3–6 feet (1–2 m) long

Place of Origin: Europe

Vegetative Characteristics: The smooth, unbranched stems of black swallowwort can either twine up through and around other vegetation or sprawl along the ground. The simple, opposite leaves are 2–4 inches (5–10 cm) long, glossy, dark green, and lance shaped with short petioles. Unlike other members of the milkweed family, the stems and leaves do not exude milky latex when broken.

Flowers and Fruit: Black swallowwort produces clusters of small, purple-black flowers with 5 triangular petals from June through September. The flowers, which can either be self- or fly-pollinated, are followed by pairs of slender green seedpods that are 2–3 inches (5–7.5 cm) long and taper to a sharp point; they contain numerous seeds tufted with silky white hairs that facilitate wind dispersal.

Germination and Regeneration: The seeds germinate in semishade or full sun and develop into perennial crowns that generate numerous shoots early in the spring. Pulling out these shoots without removing the buried crown has little effect on the plant's ability to survive and spread.

Habitat Preferences: This drought-adapted species can tolerate a wide variety of environmental conditions from full sun to dense shade. It grows in disturbed or neglected urban habitats, including residential and commercial landscapes, minimally maintained public parks and open spaces, vacant lots, urban meadows, chain-link fence lines, rock outcrops, stone walls, unmowed highway banks, and roadsides with heavy salt applications. Plants growing at the base of ornamental shrubs can become virtually impossible to remove.

Ecological Functions: Disturbance-adapted colonizer of bare ground; erosion control.

Cultural Significance: Black swallowwort was first reported in North America in the mid-1800s in Massachusetts, presumably an escape from the Harvard Botanical Garden. Its distribution in the Northeast has increased dramatically since the 1970s, and many states now list it as an invasive species. The plant is toxic and has been used in European traditional medicinal as a laxative, diuretic, emetic, and antitumor agent.

Related Species: **Pale** or **European swallowwort** (***Cynanchum rossicum*** (**Kleopov**) **Barbarich**) is similar to black swallowwort in all respects except that its flowers are pale yellow to pink-purple. The Greek physician Dioscorides included the plant in his first-century herbal compendium, *De Materia Medica*.

Black swallowwort growth habit

Black swallowwort's underground root and shoot system

Black swallowwort seedpods about to open

Black swallowwort flowers

Black swallowwort dispersing its seeds

Pale swallowwort flowers
(photo by Lou Wagner)

Achillea millifolium L.   Yarrow

SYNONYMS: milfoil, noble yarrow, staunchweed, soldier's woundwort, green arrow, nosebleed, sanguinary, yarroway, herbe militaris, knight's milfoil

LIFE FORM: herbaceous perennial; up to 3 feet (1 m) tall

PLACE OF ORIGIN: Europe, Asia, and North America

VEGETATIVE CHARACTERISTICS: Yarrow is an upright plant with hairy stems and alternate, pinnately compound leaves up to 6 inches (15 cm) long; the leaves are subdivided multiple times, giving them a finely dissected appearance. A basal rosette of leaves typically stays green through the winter; the fibrous root system is extensive but shallow.

FLOWERS AND FRUITS: Yarrow produces flat-topped, terminal clusters of white or pink flowers throughout the summer. The individual flower heads are less than 0.25 inches (6 mm) in diameter and consist of 5 "petals" (ray florets) surrounding 10 or more yellow disk florets that are pollinated by insects. At maturity the seed heads contain hundreds of tiny brown seeds that are dispersed as the plant blows back and forth in the wind. The spent flower stalks persist through the winter, making the dormant plant easy to identify.

GERMINATION AND REGENERATION: Seeds germinate readily on bare ground; established plants spread by underground rhizomes.

HABITAT PREFERENCES: Yarrow is an extremely drought-tolerant species that grows well in sunny, dry soil. It is common in minimally maintained public parks, vacant lots, rubble dumps, urban meadows, unmowed highway banks, and roadsides. In its native habitat yarrow grows in grasslands and open woodlands.

ECOLOGICAL FUNCTIONS: Tolerant of roadway salt and compacted soils; erosion control on slopes.

CULTURAL SIGNIFICANCE: Tea made from flowering yarrow plants has a long history of use in European and Chinese folk medicine for treating colds, fevers, and inflammation; the crushed leaves have been used in poultices for dressing wounds. According to Greek mythology, Achilles learned the medicinal value of yarrow from the centaur Chiron and used the plant to heal his soldiers' wounds while fighting in Troy. The plant is listed in Dioscorides' first-century herbal, *De Materia Medica*. Native Americans used yarrow, which is also native to North America, to treat numerous ailments and as a sacred sweat-lodge herb. Yarrow is widely cultivated as a drought-tolerant perennial, and horticultural selections are available in a variety of colors including red, pink, yellow, gold, and white. Landscape architecture students sometimes use the dried flower heads as trees in their models.

Stand of yarrow

Yarrow foliage

Yarrow growth habit

Yarrow in full flower

Close-up of
yarrow flower
heads

Ambrosia artemisiifolia L.   Common Ragweed

SYNONYMS: *Ambrosia elatior, A. media*, hayfever weed, wild tansy, bitter weed, hog-weed, wild wormwood

LIFE FORM: **summer annual**; up to 4 feet (1.3 m) tall

PLACE OF ORIGIN: North America

VEGETATIVE CHARACTERISTICS: The stems upright, hairy, and rough to the touch; the highly dissected, compound leaves are 2–4 inches (5–10 cm) long and give rag-weed a lacy appearance. The leaves on the lower portion of the stem are opposite; those on the upper part are mostly alternate. The plant remains green until late in the growing season.

FLOWERS AND FRUIT: The yellow-green male flowers are arranged in conspicuous terminal spikes that occupy the top one-third of the plant and resemble miniature candelabras. They produce copious amounts of highly allergenic, wind-dispersed pollen in September and October. The inconspicuous female flowers are located in leaf axils lower on the stem and produce 1-seeded fruits.

GERMINATION AND REGENERATION: The seeds germinate readily in late spring on bare ground once the temperature is consistently above 50° F (10° C). Seeds can remain viable in the soil for up to 80 years.

HABITAT PREFERENCES: Ragweed grows equally well in dry, sandy soil and heavy, moist soils with a neutral or higher pH. It is common along roadsides with heavy salt applications (which elevate soil pH); in minimally maintained public parks and residential landscapes; vacant lots and rubble dumps; small-scale pavement openings and cracks; chain-link fence lines; rock outcrops and stone walls; and lawns with compacted soil, where mowing reduces it to a stub that manages to flower despite being only 2 inches (5 cm) tall.

ECOLOGICAL FUNCTIONS: Disturbance-adapted species; tolerant of compacted soil; food for wildlife (especially birds).

CULTURAL SIGNIFICANCE: Ragweed is notorious for producing abundant pollen that causes hay fever in many people in late summer and early fall. Plants grown in a controlled environment with elevated CO_2 produced roughly a third more pollen than plants grown in ambient air, suggesting that ragweed may become a more serious allergy problem in the future (Ziska et al., 2003). The plant can accumulate lead from the soil. Philadelphia naturalist John Bartram had this to say about rag-weed in 1759: "Ye lesser ambrosia is a very troublesome weed in plantations where it hath got ahead. It is an annual & grows with corn & after harvest it shoots above ye stuble growing 3 or 4 foot high & so thick that one can hardly walk through. It tastes very bitter & if milch cows feeds upon it (for want of enough grass) their milk will taste very loathsome. It seldom grows to any head next year nor until ye field is plowed or sowed again." Ragweed has become an invasive species in parts of temperate Asia and Europe.

Ragweed foliage

Growth habit of flowering ragweed

Ragweed dominating the disturbed edge of a Connecticut bike path

Repeatedly mowed ragweed flowering at 3 to 4 inches (7.5–10 cm) tall

Male flowers of ragweed

Spiny clusters of female ragweed flowers located below male flowers

Arctium minus (Hill) Bernh.   Common Burdock

Synonyms: *Lappa minor,* clotbur, cuckoo-button, lesser burdock, velcro plant

Life Form: biennial; up to 6 feet (2 m) tall

Place of Origin: Eurasia

Vegetative Characteristics: Burdock produces a rosette of large evergreen leaves in its first year and a tall, erect, multibranched flower stalk in the second year. The alternate, heart-shaped leaves can be up to 20 inches (50 cm) long and 16 inches (40 cm) wide; their upper surface is dark green and smooth, the undersurface is light green and woolly. The stout, fleshy taproot makes the plant difficult to remove.

Flowers and Fruit: Terminal clusters of small, red-purple, thistlelike flower heads—consisting entirely of disk florets—are produced at the tips of all the braches at the end of burdock's second year. The flower heads, which are produced from July through October, are 0.75–1.5 inches (2–4 cm) wide and, following pollination by insects, develop into a bur covered with spiny, hooked bracts that attach and hold tenaciously to animal fur and human clothing.

Germination and Regeneration: Burdock seeds germinate readily in disturbed sunny or shady locations with moist soil; there is no vegetative regeneration.

Habitat Preferences: Burdock is found in a wide variety of habitats but grows best in moist, rich soil in full or half sun. It is common in minimally maintained public parks; vacant lots and rubble dump sites; the edges of emergent woodlands; the sunny borders of freshwater wetlands, ponds, and streams; and on unmowed highway banks and median strips with frequent salt applications.

Ecological Function: Disturbance-adapted colonizer of disturbed ground.

Cultural Significance: Burdock has a long history of use in traditional medicine in both Europe and China: a tea from the root is used as a "blood purifier" and to treat gout, rheumatism, and liver and kidney ailments; the seeds are used to treat various skin eruptions. The Greek physician Dioscorides included the plant in first-century herbal, *De Materia Medica.* The peeled stems and root are edible when boiled; indeed, the Japanese vegetable *gobo* comes from domesticated varieties of burdock's close relative, *Arctium lappa.* Burdock was an early arrival in North America and is listed in Josselyn's *New-England's Rarities,* published in 1672. In the Shakespeare play *As You Like It,* burdock is a prominent metaphor in a conversation between Rosalind and Celia, who notes with sadness, "O how full of briars is this work-a-day world." To which the cheerful Rosalind responds, "They are but burs, cousin, thrown upon thee in holiday foolery. If we walk not in the trodden paths, our very petticoats will catch them." The melancholy Celia replies, "I could shake them off my coat; these burs are in my heart." And finally, the Swiss engineer George de Mestral came up with the idea for Velcro® when he studied burdock burs under the microscope after removing them from his dog following a hike in the Alps.

Burdock foliage at the end of its first growing season

Burdock growth habit

Burdock at the end of its second growing season

Burdock flower heads

Developing burdock seed heads can be used as a brooch

Artemesia vulgaris L.   *Mugwort*

SYNONYMS: chrysanthemum weed, common wormwood

LIFE FORM: herbaceous perennial; up to 6 feet (2 m) tall

PLACE OF ORIGIN: Eurasia

VEGETATIVE CHARACTERISTICS: Mugwort is a tall, upright plant that develops thick, woody stems by late in the season. Its leaves are alternate, 2–4 inches (5–10 cm) long, and deeply dissected with sharp-pointed lobes. The upper surface of the leaf is smooth; the underside is covered with woolly white hairs, giving the plant a gray-green appearance, especially after it begins to flower in August. The foliage gives off a pungent odor when crushed.

FLOWERS AND FRUIT: Mugwort produces a feathery terminal inflorescence consisting of numerous tiny, greenish yellow, wind-pollinated flower heads—made up entirely of disk florets—in late summer and fall. The weight of the developing fruits typically causes the plant to arch over at the end of the growing season.

GERMINATION AND REGENERATION: Mugwort seeds are quite small (2 mm long) and seem to be dispersed mainly by wind and also by water. Buried seeds remain viable in the soil for many years. Established plants spread vigorously by rhizomes.

HABITAT PREFERENCES: Mugwort is the quintessential urban weed. It thrives on disturbed, compacted soil with high pH and is common in minimally maintained public parks, vacant lots, rubble dumps, soil stockpiles, small pavement openings and cracks, and along railroad tracks. It recently has begun to spread along highway margins and median strips in suburban and rural areas, most likely in response to heavy applications of road salt.

ECOLOGICAL FUNCTIONS: Phytoremediation in degraded urban landscapes (absorbs the heavy metals zinc, copper, lead, and cadmium and binds them to organic matter); tolerant of road salt and compacted soil; erosion control on slopes; soil building on degraded soil.

CULTURAL SIGNIFICANCE: Mugwort was used to flavor beer before hops took over that role (hence its common name). In traditional Chinese medicine, the compressed, dried leaves are burned on the skin to stimulate acupuncture points (moxa) and to treat rheumatism. In European traditional medicine, a tea made from mugwort leaves was used to treat epilepsy; menstrual, menopausal, and gastrointestinal problems; and to increase the flow of urine and stimulate the appetite. European and Asian cuisines use the young leaves to flavor a variety of traditional dishes. The plant was introduced into North America at an early date, undoubtedly for medicinal purposes. Mugwort pollen is a potent allergen and can be a major cause of fall hay fever.

Stand of mugwort in May (note silvery leaf undersides)

Mugwort leaf shape is highly variable

Mugwort dominating a neglected urban sidewalk

Mugwort foliage just coming up in early spring

Mugwort in flower in September

Bidens frondosa L.   Devil's Beggarticks

SYNONYMS: sticktights, devil's bootjack, cuckold, beggarticks, bur-marigold

LIFE FORM: summer annual; up to 5 feet (1.5 m) tall

PLACE OF ORIGIN: eastern North America

VEGETATIVE CHARACTERISTICS: Devil's beggarticks is an upright, loosely branched plant with 4-angled, often purplish stems. The opposite, compound leaves have 3–5 large, lance-shaped leaflets up to a foot (30 cm) long with serrated edges.

FLOWERS AND FRUIT: Devil's beggarticks produces green and yellow flower heads, about 1 inch (2.5 cm) wide, from August through October; they are mainly self-pollinated. The narrow, leaflike bracts that surround the base of the flower head are conspicuously longer than the orange-yellow petals (ray florets) that surround the yellow-brown disk florets. The brown fruits (achenes), which mature in late October, are about 0.5 inch (1.2 cm) long with 2 barbed spines that look like an old-fashioned boot jack and stick tenaciously to clothing.

GERMINATION AND REGENERATION: The seeds are dispersed by animals and humans and germinate under a wide variety of conditions.

HABITAT PREFERENCES: Devil's beggarticks grows in a wide variety of habitats ranging from dry and shady to moist and sunny; it is common in minimally maintained public parks; woodlands that develop on abandoned open space; the edges of freshwater wetlands, ponds, and streams; drainage ditches; and landscape planting beds.

ECOLOGICAL FUNCTIONS: Tolerant of compacted soil; stream and river bank stabilization.

CULTURAL SIGNIFICANCE: Native Americans used a tea made from the leaves to expel worms and chewed the leaves for sore throat. The Shakers sold the plant as an expectorant and to treat heart palpitations and "uterine derangement," and to induce sweating, menstruation, and urination.

Devil's beggarticks foliage

Devil's beggarticks growth habit in flower

Devil's beggarticks typically grows in moist, shady conditions

Devil's beggarticks flower heads

Devil's beggarticks developing seed head

Centaurea biebersteinii DC   Spotted Knapweed

SYNONYMS: *Centaurea maculosa*, star thistle

LIFE FORM: biennial or perennial; 1–4 feet (30–130 cm) tall

PLACE OF ORIGIN: Europe

VEGETATIVE CHARACTERISTICS: Spotted knapweed forms a basal rosette of deeply lobed, gray-green leaves about 6 inches (15 cm) long during its first year. In its second year it produces slender, wiry stems and gray-green leaves that are pinnately dissected with numerous lobes; the leaves near the top of the plant are tiny and narrow.

FLOWERS AND FRUIT: Thistlelike, pink-purple flower heads that are about 0.5–1 inch (1.5–2.5 cm) wide are produced at the ends of all the branches from mid- to late summer. The heads consist entirely of tubular disk florets, which are pollinated by a variety of insects (especially honeybees). The dry fruits (achenes) have short bristles that facilitate animal dispersal but are also dispersed by wind.

GERMINATION AND REGENERATION: The seeds germinate readily in fall or spring in dry, sandy soil in full sun; mature plants produce short rhizomes which give rise to new shoots.

HABITAT PREFERENCES: Spotted knapweed grows well in disturbed sites with dry, low-fertility soils. It is common in vacant lots and along roadways and railroad tracks. The plant is highly salt tolerant and common near the coast.

ECOLOGICAL FUNCTIONS: Disturbance-adapted colonizer of bare ground; tolerant of compacted soil. Spotted knapweed produces allelopathic chemicals that act as herbicides to suppress the growth of nearby plants of a different species, one of the reasons for its success as an invasive species.

CULTURAL SIGNIFICANCE: Spotted knapweed is a fairly recent, unintentional introduction into North America, having first been recorded in the 1890s. It has since become a major problem in the rangelands of Idaho, Montana, and eastern Oregon and Washington. Federal and state agencies have spent millions of dollars trying to control spotted knapweed, and many states list it as an invasive species.

Spotted knapweed in a typical urban habitat

Spotted knapweed growth habit

Spotted knapweed rosette foliage

Spotted knapweed in flower

Spotted knapweed flower head

Cichorium intybus L.   Chicory

SYNONYMS: succory, blue sailors, ragged sailors, blue daisy, coffee-weed, wild endive

LIFE FORM: herbaceous perennial; up to 5 feet (1.7 m) tall

PLACE OF ORIGIN: Eurasia

VEGETATIVE CHARACTERISTICS: Chicory forms a basal rosette of dark green, lance-shaped leaves with coarsely toothed margins that resemble those of dandelion. The basal leaves are 3–9 inches (8–25 cm) long by 1–3 inches (2–7 cm) wide; the upper leaves are much smaller and their bases clasp the stem. Mowed plants are much shorter than those that are allowed to grow freely. Chicory forms a deep, perennial taproot, and all parts of the plant exude milky sap when broken.

FLOWERS AND FRUIT: Chicory's tall flower stalks emerge from basal rosettes from July through October; the bright blue (and occasionally pink) flower heads, which are about 1.5 inches (4 cm) across, open in the morning and close by afternoon. The individual florets that make up the flower head are all of the ray type with a single, attached blue "petal," and are commonly pollinated by bees.

GERMINATION AND REGENERATION: Birds eat and disperse the dried fruit, and the seeds germinate in a variety of sunny sites, especially in pavement cracks and along roadside edges; established plants sprout back from the base in early spring.

HABITAT PREFERENCES: Chicory is a common roadside plant that grows well in sunny, dry habitats and soils with a bit of limestone or an elevated pH. In the urban environment it is common in minimally maintained public parks, the margins of neglected residential and commercial landscapes, vacant lots, rubble dumps, pavement cracks, chain-link fence lines, unmowed highway banks and median strips, and railroad rights-of-way.

ECOLOGICAL FUNCTIONS: Tolerant of roadway salt and compacted soil; food and habitat for wildlife; erosion control on slopes; soil building on degraded land.

CULTURAL SIGNIFICANCE: Domesticated varieties of chicory, known as Belgium endive, have distinctive spoon-shaped, pale white leaves that are produced by growing year-old plants in the dark. This technique, known as blanching, prevents the new leaves from turning green and makes them less bitter than they would be if they were grown in the sun. Radicchio is a red-leaved variety of chicory that has been cultivated in the Mediterranean region for thousands of years. The root of chicory, when roasted and ground, has long been used as a coffee additive or substitute, especially during times of scarcity. The root has also been used in traditional European medicine to treat liver problems, gout, sour stomach, and rheumatism. Chicory has more recently been evaluated as a forage species for feeding livestock; for phytomining and phytoremediation purposes (based on its capacity to absorb heavy metals); and as a commercial source of inulin, which is used as a sweetener in the food industry. The leafy salad greens escarole and curly endive (*Cichorium endivia* L.) are often confused with chicory.

Chicory in full flower

Chicory foliage rosettes before flowering

Chicory growth habit as its flower stalks are expanding

Stand of chicory in a bed of gravel

Close-up of chicory flower heads

Cirsium vulgare (Savi) Tenore   Bull Thistle

Synonyms: *Carduus vulgaris; Cirsium lanceolatum;* spear thistle

Life Form: evergreen biennial; up to 6 feet (2 m) tall

Place of Origin: Eurasia and North Africa

Vegetative Characteristics: During its first year bull thistle forms a rosette of spiny, dark green leaves about 8 inches (20 cm) long that persists through the winter. The following spring the rosette "bolts" and produces a tall, upright stalk. The leaves are alternate, with sharp spines on their lobes and bases that fuse with the stem, giving it a winged or fluted appearance.

Flowers and Fruit: Bull thistle produces reddish purple flower heads at the ends of the stalks from June through October. These heads, which are about 1.5 inches (4 cm) wide, consist entirely of disk florets; their bases are covered with sharp-tipped bracts. Following insect- or self-pollination, a spiny capsule develops that is filled with small seeds, each topped with a downy pappus.

Germination and Regeneration: The wind-dispersed seeds germinate readily in full sun in a variety of habitats and soil types. There is no form of vegetative regeneration.

Habitat Preferences: Bull thistle grows best in moist, rich soil in full sun. In the urban environment it is common in vacant lots and rubble dumps; abandoned grasslands; along the margins of freshwater wetlands, ponds, and streams; on unmowed highway banks and median strips; and along railroad tracks.

Ecological Function: Disturbance-adapted colonizer of bare ground.

Cultural Significance: When fully mature, this biennial is a dramatic presence in the landscape. The young, elongating stems are edible when peeled and boiled. The plant was inadvertently introduced into North America sometime in the 1700s as a contaminant in seed or hay. Thistles have always had image problems, dating back to Old Testament days when God banished Adam from the Garden of Eden with the words, "Cursed is the ground because of you . . . It will produce thornes and thistles for you" (Genesis 3, verses 17–18).

Related Species: Canada thistle (***Cirsium arvense* (L.) Scop.**) is a dioecious perennial species that grows to about 4 feet (1.3 m) tall. It produces numerous pink-purple flower heads about an inch (2.5 cm) across throughout the summer and begins seeding prolifically in August. It grows best in sunny sites with good soil and produces vigorous new shoots from deep, creeping roots, eventually forming large colonies. Canada thistle is difficult to eradicate and was one of the first plants put on the U.S. government's noxious weed list. The plant has a long history of medicinal use in Europe as a tonic, a diuretic, and an astringent for treating skin sores and rashes.

Bull thistle flower and seed heads

Bull thistle foliage

Bull thistle
rosette after
its first year of
growth

Canada thistle growth habit

Bull thistle at the end of its
life span

Canada
thistle
flower
heads

Conyza canadensis (L.) Cronq.   Horseweed

SYNONYMS: *Erigeron canadensis*, Canada fleabane, mare's tail, butter weed

LIFE FORM: winter or summer annual; up to 6 feet (2 m) tall

PLACE OF ORIGIN: North America

VEGETATIVE CHARACTERISTICS: Horseweed's tall, hairy stems do not branch until July or August, when it comes into flower. The alternate leaves are about 4 inches (10 cm) long by 0.5 inch (1.5 cm) wide and are crowded along the stems. Seedlings often form a basal rosette of leaves that deteriorates as the stems begin to elongate in early summer.

FLOWERS AND FRUIT: Horseweed produces masses of tiny white flowers at the ends of all the upper branches from July through October. The individual flower heads consist of numerous small "petals" (ray florets) surrounding the yellow disk florets, which can be self- or insect-pollinated. The weight of the developing seeds gives the plant its top-heavy appearance and often causes it to bend over or break.

GERMINATION AND REGENERATION: The tiny horseweed seeds are topped with whitish bristles that facilitate wind dispersal; they can germinate on a wide variety of disturbed urban sites.

HABITAT PREFERENCES: Horseweed grows well in dry, exposed sites in full sun. It is common in minimally maintained public parks and residential landscapes, vacant lots, rubble dumps, abandoned grasslands, rock outcrops, stone walls, fence lines, small pavement openings, railroad rights-of-way, and unmowed highway banks and median strips.

ECOLOGICAL FUNCTION: Disturbance-adapted colonizer of bare ground.

CULTURAL SIGNIFICANCE: Native Americans used horseweed to treat diarrhea, and the Shakers sold it to treat kidney stones and a variety of bowel and urination problems. Horseweed has become a major invasive species in temperate Asia and Europe.

Stand of horseweed in Boston

Horseweed growing along railroad tracks

Horseweed getting ready to flower

Close-up of horseweed flower heads

Horseweed in full seed by a roadside

Erechtites hieracifolia (L.) Raf. ex DC.   Fireweed

SYNONYMS: pilewort, American burnweed

LIFE FORM: **summer annual**; up to 6 feet (2 m) tall

PLACE OF ORIGIN: eastern North America

VEGETATIVE CHARACTERISTICS: Fireweed has smooth, unbranched, bright green stems and leaves. The leaves are alternate, oblong to lance shaped, pinnately lobed, and have sharp irregular teeth along their margins; they can be up to 6 inches (15 cm) long by 2 inches (5 cm) wide.

FLOWERS AND FRUIT: Fireweed flower heads have a tight, tubular form, and the individual disk florets barely extend above the upright green bracts. Dozens of these composite flowers are typically arranged into a terminal, flat-topped cluster. The flower heads are self- or insect-pollinated and quickly mature into "puffball" seed heads composed of small black fruits topped with a cluster of slender, soft white hairs (the pappus).

GERMINATION AND REGENERATION: The wind-dispersed seeds are famous for their ability to germinate on land that has been recently burned, but fireweed also grows on sites that have been disturbed by other agents (see below). Less well appreciated is the fact that the seeds can germinate after years of burial when some form of physical disturbance brings them to the surface.

HABITAT PREFERENCES: In the urban environment fireweed grows in a variety of disturbed sites in full sun, typically as a single plant or in small groups, but occasionally in large patches. It is common in small pavement openings and pavement cracks, along the margins of walls and chain-link fences, in minimally maintained landscape plantings, and along railroad tracks.

ECOLOGICAL FUNCTION: Disturbance-adapted colonizer of bare ground.

CULTURAL SIGNIFICANCE: Native Americans and early European settlers used fireweed to treat hemorrhoids (piles), and the Shakers sold an extract of the plant to treat "diseases of the mucous tissues of the lungs, stomach and bowels." John Bartram published a remarkably detailed account of the ecology of fireweed in 1759: "We have another weed called Cotton groundsel [*Erechtites*] which grows with us 6 or 7 foot high & ye stalk at bottom near as thick as my wrist. In our new cleared land after ye first plowing in ye spring or in our marshes ye year after they are drained it grows there all over so close that there is no passing along without breaking it down to walk or ride through it but in ould fields or medows there is not one stalk to be seen now. If we put ye question how comes this to grow so prodigiously on ye new land plowed ground & perhaps not one root growing within several miles ye answer is very ready. . . . One day when ye sun shined bright a little after its meridian my Billy [Bartram's son] was looking up at it when he discovered an innumerable quantity of downey motes floating in ye air between him & ye sun. . . . Some lowered & fell into my garden where we observed every particular detachment of down spread in 4 or 5 rays with a seed of ye grounsel in its center. How far these was carried by that breeze cant be known but I think thay must have come near five miles from a meadow to reach my garden."

Fireweed growth habit in fall

Fireweed foliage

Fireweed flower heads

Fireweed dispersing its seeds

Fireweed seeds ready for dispersal

Erigeron annuus (L.) Pers.   Annual Fleabane

Synonyms: daisy fleabane, white top

Life Form: **winter or summer annual**; up to 3–4 feet (1–1.3 m) tall

Place of Origin: eastern North America

Vegetative Characteristics: Annual fleabane is an upright plant with ridged, hairy stems and alternate leaves with toothed margins. Leaves at the base of the stem are up to 4 inches (10 cm) long, elliptical or egg shaped with distinct petioles; those at the top are linear or lance shaped with short or no petioles.

Flowers and Fruit: Plants produce terminal clusters of compound flower heads, about 0.5 inch (1–1.5 cm) in diameter, made up of numerous thin, white "petals" (ray florets) surrounding a yellow center (disk florets). The flowers are commonly visited by insects but recent research indicates that the small, wind-dispersed seeds are produced asexually and are genetically identical with the plant that produced them—a form of clonal reproduction known as *apomixis*.

Germination and Regeneration: The seeds germinate in spring in sunny, disturbed sites.

Habitat Preferences: Annual fleabane grows best in fertile soil in full sun but also tolerates dry soil. In the urban environment it is common along the margins of minimally maintained public parks; in vacant lots, rubble dumps, abandoned lawns, urban meadows, and unmowed highway banks and median strips; and along railroad tracks.

Ecological Function: Disturbance-adapted colonizer of bare ground.

Cultural Significance: Native Americans used annual fleabane to treat a variety of ailments. Burning it is said to drive away fleas and gnats (hence the common name), but there is no evidence that this is true. The species has become weedy in Europe, reversing the typical Europe to North America scenario.

Daisy fleabane growth habit

Field of daisy fleabane

Daisy fleabane in full bloom

Daisy fleabane growing between
sidewalk and street

Daisy fleabane
flower head

Galinsoga quadriradiata Cav.   Hairy Galinsoga

SYNONYMS: *Galinsoga ciliata*, quick weed, gallant soldier, galinsoga

LIFE FORM: summer annual; up to 2 feet (70 cm) tall

PLACE OF ORIGIN: Central America

VEGETATIVE CHARACTERISTICS: Hairy galinsoga is densely branched and completely covered with coarse hairs; the leaves are opposite, oval to triangular, coarsely toothed, and 1–3 inches (2.5–7.5 inches) long. The root system is fibrous, shallow, and relatively easy to pull up.

FLOWERS AND FRUIT: The tiny flower heads of hairy galinsoga, which are about 0.25 inch (8 mm) wide, are produced from June through frost. They consist of 5 white "petals" (ray florets) with notched tips surrounding a central core of yellow disk florets. The florets can be either insect- or self-pollinated and produce tiny seeds (achenes) with a tuft of papery scales that facilitate their dispersal.

GERMINATION AND REGENERATION: Galinsoga seeds germinate throughout the summer on disturbed soil. A single plant can produce as many as 3 generations in a season, making for a population explosion in late summer. Buried seeds retain their viability in the soil for many years.

HABITAT PREFERENCES: In full sun and fertile soil hairy galinsoga can become quite large and robust. In the urban environment it is found in neglected residential and commercial landscapes, minimally maintained public parks and open spaces, vacant lots, rubble dumps, small pavement openings, and sidewalk cracks. The plant is noteworthy for its ability to complete its life cycle from germination to seed production in as little as 3–4 weeks.

ECOLOGICAL FUNCTION: Disturbance-adapted colonizer of bare ground.

CULTURAL SIGNIFICANCE: Hairy galinsoga arrived in the Northeast in the 1920s, having migrated up from Central America and through the Southeast; it is now ubiquitous throughout eastern North America. The plant has also become a common weed in Asia, where young plants are eaten in the spring.

Hairy galinsoga growth habit

Hairy galinsoga in its typical urban habitat

Hairy galinsoga foliage

Hairy galinsoga flower heads

Close-up of hairy galinsoga flower heads

Hieracium sabaudum L.   New England Hawkweed

Synonyms: Even experts have trouble distinguishing this highly variable European species from another European species, *Hieracium lachenalii,* and a native North American species, *Hieracium canadense.*

Life Form: herbaceous perennial; up to 3 feet (1m) tall

Place of Origin: Eurasia

Vegetative Features: New England hawkweed produces erect stems with alternate, dark green, lance-shaped leaves that are 3–5 inches (7–12 cm) long. The leaves have sharp-pointed teeth scattered along the margins and are hairy on the underside. All parts of the plant exude milky sap when broken. Unlike many other hawkweeds, this upright species does not form a basal rosette.

Flowers and Fruit: Conspicuous bright yellow flower heads, similar to those of dandelion, are produced from August through October. The flower heads, which are at the ends of long, highly branched stems, are about an inch (2–3 cm) wide and consist entirely of ray florets. They can be self- or insect-pollinated and develop into round, "puffball" fruiting heads about 0.75 inch (2 cm) wide that are composed of numerous small, wind-dispersed seeds.

Germination and Regeneration: Seeds germinate best in moist sites in partial sun. Established plants produce basal buds that can give rise to dense clusters of new stems. Unlike many other hawkweeds, this species is not rhizomatous.

Habitat Preferences: New England hawkweed grows best in dry, shady sites with compacted or sandy soil, but it can also tolerate full sun. In the Northeast it is much more common in the urban environment than in the countryside. It is found especially along the shady margins of disturbed or emergent woodlands, in the rain shadow of buildings, and around the base of rock outcrops.

Ecological Functions: Tolerant of road salt and compacted soil; erosion control on slopes; food for insect pollinators.

Cultural Significance: New England hawkweed appears to have been introduced into the Boston area in the late 1800s. It now occurs throughout the Northeast and mid-Atlantic regions.

Related Species: Field hawkweed (*Hieracium caespitosum* Dumort. [formerly known as *H. pratense*]) is a mat-forming European species that spreads by rhizomes. It produces rosettes of lance-shaped leaves that are conspicuously hairy on both sides and 2–10 inches (5–25 cm) long. In June and July the plant produces a hairy, leafless flower stalk 1–3 feet (30–90 cm) tall that is topped with 4 or more yellow, dandelion-like flowers that are about an inch (2.5 cm) wide. Field hawkweed is common in the dry, nutrient-poor, acid soils of urban meadows and gravelly roadsides.

New England hawkweed growth habit in early summer

New England hawkweed foliage in spring

New England hawkweed in flower at the edge of a woodland path

New England hawkweed flower heads

New England hawkweed flower and seed heads

Field hawkweed in flower

Lactuca serriola L.   Prickly Lettuce

Synonyms: *Lactuca scariola*, compass plant

Life Form: **annual** or **biennial**; up to 6 feet (2 m) tall

Place of Origin: Europe

Vegetative Characteristics: Prickly lettuce produces a basal rosette of pale green leaves that typically gives rise to a single hollow, upright, green to white stem. Leaves are alternate, 2–10 inches (5–25 cm) long, and usually (but not always) deeply lobed with rounded, half-moon-shaped sinuses; the upper surface of the leaf is smooth, but a single row of stiff prickles is conspicuous along the midrib of the lower surface. The name "compass plant" comes from the fact that the leaves are typically turned on edge and oriented vertically (rather than horizontally) toward the sun. All parts of the plant exude milky sap when broken.

Flowers and Fruit: The yellow flower heads are less than 0.25 inch (1 cm) wide and are grouped into a large, pyramidal panicle that terminates the main stem. The flower heads consist entirely of ray florets and are produced from July through September; they can either be self- or insect-pollinated. Seeds develop rapidly, and each has a feathery pappus to aid in its dispersal by wind.

Germination and Regeneration: The seeds germinate readily in a variety of soils and exposures.

Habitat Preferences: Prickly lettuce grows best in nutrient-rich soils and full sun but can tolerate dry sites with poor soil. In the urban environment it is common in neglected residential and commercial landscapes, minimally maintained public parks, unmowed highway banks and median strips, and small pavement openings and cracks.

Ecological Function: Phytoremediation in degraded urban landscapes by absorbing heavy metals (mainly zinc and cadmium) and binding them to organic matter.

Cultural Significance: The wild ancestor of cultivated lettuce, this plant has been used in traditional European medicine as a diuretic and to stimulate the flow of milk in nursing mothers. The young shoots can be eaten but are not recommended.

Prickly lettuce with entire (unlobed) leaves

Prickly lettuce with lobed leaves (note orientation perpendicular to the sun)

Prickly lettuce in its typical urban habitat between a sidewalk and a wall

Note the stiff prickles along the underside of prickly lettuce leaf

Prickly lettuce flower heads

Prickly lettuce seed heads

Leontodon autumnalis L.   Fall Dandelion

SYNONYM: fall hawkbit

LIFE FORM: herbaceous perennial; up to 2 feet (60 cm) tall

PLACE OF ORIGIN: Eurasia

VEGETATIVE CHARACTERISTICS: Fall dandelion produces a rosette of smooth, narrow leaves up to a foot (30 cm) long with deep, irregular lobes that have rounded points. It has a fibrous root system, and all parts of the plant exude milky latex when broken.

FLOWERS AND FRUIT: Fall dandelion produces bright yellow flower heads from late June through October that resemble those of dandelion but are somewhat smaller, only about an inch (2.5 cm) across. The flower heads consist entirely of ray florets and are produced at the ends of thin, forked stalks that bear a few small scales. Following pollination by insects, brown "puffball" fruiting heads quickly form; each head is composed of numerous small, wind-dispersed seeds topped by a feathery pappus.

GERMINATION AND REGENERATION: Seeds germinate readily in a variety of disturbed habitats; established plants emerge from a perennial crown and can also produce new plants from short rhizomes.

HABITAT PREFERENCES: Fall dandelion is highly tolerant of mowing and is abundant in compacted, minimally maintained lawns and ball fields. It also grows in vacant lots, rubble dumps, small pavement openings and cracks, and roadsides. In its native habitat fall dandelion grows in grasslands and on cliffs.

ECOLOGICAL FUNCTION: Tolerant of roadway salt and compacted soil.

CULTURAL SIGNIFICANCE: In 1847, Darlington noted that the plant was "especially abundant in New England where it infests grass plots." His statement is still accurate today.

SIMILAR SPECIES: While the flowers of dandelion (*Taraxacum officinale*) are somewhat similar, they are produced mainly in spring and early summer rather than from late June into October. Dandelion seed heads are also larger, whiter, and more symmetrical than those of fall dandelion, which are brown and irregular.

Fall dandelion foliage

Fall dandelion flourishing in a dry lawn

Contrasting foliage of dandelion (*left*) and fall dandelion (*right*)

Fall dandelion in flower in late summer

Fall dandelion flower heads

Leucanthemum vulgare Lam.   Oxeye Daisy

Synonyms: *Chrysanthemum leucanthemum*, white daisy, white weed, marguerite, snow-in-June

Life Form: herbaceous perennial; up to 2 feet (70 cm) tall

Place of Origin: Europe

Vegetative Characteristics: In early spring, oxeye daisy produces a basal rosette of hairless, dark green, alternate leaves—up to 6 inches (15 cm) long—with rounded lobes and long petioles. As the flower stalks elongate in late spring, the plant produces sessile leaves that become progressively smaller, narrower, and less lobed toward the top.

Flowers and Fruit: This species produces classic daisy "flowers" at the ends of long stalks from June through July. The insect-pollinated flower heads are 1–2 inches (3–5 cm) across with 20–30 white "petals" (ray florets) surrounding a cluster of yellow disk florets. The fruits (achenes) are tiny and lack the bristles that facilitate wind dispersal in many other members of the Asteraceae.

Germination and Regeneration: The small, dark-brown seeds are dispersed by wind and birds. Buried seeds retain their viability in the soil for many years. Established plants can form large clumps that increase in size through the growth of rhizomes.

Habitat Preferences: Oxeye daisy is highly drought tolerant and grows best in full sun. It is common in compacted lawns, minimally maintained public parks, vacant lots, rubble dumps, urban meadows, rock outcrops, stone walls, unmowed highway banks and median strips, and railroad rights-of-way.

Ecological Functions: Tolerant of roadway salt and compacted soil; food for wildlife; soil building on degraded land; erosion control on slopes.

Cultural Significance: Most people assume that this common, attractive species is native to North America, but it was introduced from Europe in the early 1700s, probably as a contaminant in hay brought over to feed livestock. The plant has a long tradition of medicinal use in Europe to treat whooping cough, asthma, and nervous excitability. Philadelphia naturalist John Bartram was inspired to become a botanist when he stopped in the middle of plowing his fields one day to look closely at an oxeye daisy: "What a shame, said my mind, or something that inspired my mind, that thee shouldst have employed so many years in tilling the earth and destroying so many flowers and plants, without being acquainted with their structures and their uses!"

Stand of
oxeye daisy
in a degraded
urban meadow

Oxeye daisy growth habit
just before flowering

Oxeye daisy foliage

Oxeye daisy flower heads

Close-up of oxeye daisy flower head

Senecio vulgaris L.   Common Groundsel

SYNONYMS: ground glutton, chickenweed, grimsel, ragwort

LIFE FORM: winter or **summer annual**; up to 18 inches (45 cm) tall

PLACE OF ORIGIN: Eurasia

VEGETATIVE CHARACTERISTICS: Groundsel produces smooth green stems that are initially erect but as they branch they bend over and form large, spreading clumps. The mat-green leaves are alternate, smooth, have irregularly toothed or deeply lobed margins, and are 6–10 inches (15–25 cm) long with a semi-succulent texture.

FLOWERS AND FRUIT: The terminal clusters of flower heads are about 0.5 inch (1 cm) wide and consist entirely of yellow disk florets (no ray florets); they are mainly self-pollinated and quickly mature into small brown seeds topped with a fluffy white pappus that aids wind dispersal. The plant is typically in flower and fruit simultaneously.

GERMINATION AND REGENERATION: The seeds germinate in both spring and fall, when the weather is cool. Seeds can survive burial in the soil for many years.

HABITAT PREFERENCES: Groundsel grows best in moist, nutrient-rich soils in spring and fall. It is common in a wide variety of urban habitats, including small pavement openings, stone walls, rock outcrops, vacant lots, waste dumps, moist lawn areas, and ornamental landscape plantings.

ECOLOGICAL FUNCTION: Disturbance-adapted colonizer of bare ground.

CULTURAL SIGNIFICANCE: Groundsel, collected in spring, has a long history of medicinal use in Europe as a diuretic and a purgative. The plant was an early arrival in North America because Josselyn listed it in *New-England's Rarities* (1672) under the category: "Of such plants as have sprung up since the English planted and kept cattle in New England." It contains alkaloids that can poison livestock.

RELATED SPECIES: Pineapple weed (*Matricaria discoidea* DC.) is a summer annual up to 18 inches (45 cm) tall that grows best in full sun and is an indicator of sandy or compacted soils; its stems are smooth and green and can be either erect or prostrate depending on growing conditions. According to Hutchinson (1948), "The more it is trodden on the better it seems to thrive." The shiny, bright green leaves are finely dissected, alternate, and 0.5–2 inches (1–5 cm) long. All of its parts have a distinct pineapple-like scent when crushed. The greenish yellow flower heads, which are 0.25–0.5 inch (5–10 mm) wide, look like cone-shaped buttons and lack ray florets; they are produced from May through September; the tiny brown fruits lack a feathery pappus. Pineapple weed seems to be native to both northeastern Asia and northwestern North America. A tea made from the leaves has been used in traditional medicine for stomachaches and colds.

Groundsel growth habit

Groundsel flower and seed heads

Close-up of groundsel flower heads

Pineapple weed growth habit

Pineapple weed about 4 inches
(10 cm) tall

Close-up of pineapple weed flower heads

Solidago canadensis L.   Canada Goldenrod

Synonyms: yellow-weed, common goldenrod

Life Form: herbaceous perennial, up to 5 feet (1.5 m) tall

Place of Origin: eastern North America

Vegetative Characteristics: Canada goldenrod is a tall, erect plant with smooth or slightly hairy stems and alternate, linear to lance-shaped leaves that are tapered at both ends. The leaves, which are up to 6 inches (15 cm) long, become smaller toward the top of the stem and have margins that are either smooth or serrated. Canada goldenrod is highly variable and can be difficult to distinguish from closely related species, especially tall goldenrod (*S. altissima*) and late goldenrod (*S. gigantea*).

Flowers and Fruit: Canada goldenrod produces conspicuous yellow flowers arranged in curved, pyramidal clusters from August through October. The individual flower heads consist of small yellow ray florets surrounding equally small yellow disk florets, which produce sticky pollen that attracts insects. Common pollinators include honeybees, bumblebees, various beetles, and butterflies. The tiny seeds have a few short bristles at one end that facilitate wind dispersal.

Germination and Regeneration: The seeds germinate in a variety of sunny sites with moist or dry soil. Creeping rhizomes grow out from the main stem in the late summer and fall and sprout the following spring to form a cluster of genetically identical stems. Over time and under the right conditions, large colonies can develop.

Habitat Preferences: Canada goldenrod grows best in well-drained soil and full sun. In the urban environment it is commonly found in neglected residential and commercial landscapes, minimally maintained public parks, vacant lots, rubble dump sites, abandoned grasslands and meadows, unmowed highway banks, drainage ditches, and railroad rights-of-way.

Ecological Functions: Disturbance-adapted colonizer of bare ground; food for a wide variety of insect pollinators and herbivores; erosion control on slopes.

Cultural Significance: The conspicuous, insect-pollinated goldenrods are often blamed for the allergies caused by the much less conspicuous, wind-pollinated ragweed, which blooms at the same time. Native Americans used all parts of the goldenrod plant for a variety of medicinal purposes. Thomas Edison experimented with goldenrod sap to produce rubber but was unable to develop a commercial product. Canada goldenrod has become a major invasive species in temperate Asia and Europe.

Canada goldenrod growth habit

Canada goldenrod foliage before flowering

Canada goldenrod rhizomes

Canada goldenrod in full flower

Close-up of Canada goldenrod flower heads

Sonchus oleraceus L.   Annual Sowthistle

SYNONYMS: common sowthistle, milk thistle, hare's lettuce, hare's thistle

LIFE FORM: summer annual; up to 6 feet (2 m) tall

PLACE OF ORIGIN: Europe

VEGETATIVE CHARACTERISTICS: Annual sowthistle is an unbranched plant with smooth, erect stems and foliage that has a waxy bluish appearance. The deeply lobed leaves are alternately arranged, up to a foot (30 cm) long, and have margins that are irregularly toothed and slightly prickly; leaf lobes near the base of the petiole seem to clasp or wrap around the stem. All parts of the plant exude milky white sap when broken.

FLOWERS AND FRUIT: Terminal clusters of pale yellow flower heads 0.5–1 inch (1.2–2.5 cm) in diameter are produced from July through October. The flower heads consist entirely of ray florets and, following self- or insect-pollination, develop into tiny white "puffballs" consisting of brown seeds with a feathery pappus that facilitates wind dispersal.

GERMINATION AND REGENERATION: Seeds germinate in a wide variety of conditions from sun to shade and moist to dry.

HABITAT PREFERENCES: Annual sowthistle grows best in rich soil in full sun. It is common in neglected residential and commercial landscapes, minimally maintained public parks, small pavement openings and cracks, vacant lots, and rubble dump sites.

ECOLOGICAL FUNCTION: Disturbance-adapted colonizer of bare ground.

CULTURAL SIGNIFICANCE: Dioscorides included annual sowthistle in his first-century herbal, *De Materia Medica*. The species name, *oleraceus*, means "of the vegetable garden" in Latin, and indeed, the plant is an edible potherb in early spring. Rabbits seem to be particularly fond of common sowthistle, which is also known as hare's lettuce. The plant was an early arrival in North America and appears in Josselyn's in *New-England's Rarities* (1672) under the category: "Of such plants as have sprung up since the English planted and kept cattle in New England."

RELATED SPECIES: The leaves of **perennial sowthistle (*Sonchus arvensis* L.)** resemble those of annual sowthistle, but the perennial plant grows much taller—up to 4.5 feet (1.5 m)—and sends up new shoots from persistent rhizomes. In July and August it produces bright yellow, dandelion-like flower heads that are nearly twice as big as those of annual sowthistle—up to 2 inches (5 cm) wide. It seems to be much more common in the countryside than in the city. The young shoots are edible in spring.

Annual sowthistle growth habit

Annual sowthistle leaf

Mature seed head of annual sowthistle

Annual sowthistle flower and
developing seed heads

Close-up of
perennial
sowthistle
flower head

Perennial sowthistle in flower

Symphyotrichum pilosum (Willd.) Nesom   White Heath Aster

SYNONYMS: *Aster pilosus*, frost aster, awl aster

LIFE FORM: herbaceous perennial; up to 3 feet (1 m) tall

PLACE OF ORIGIN: eastern and central North America

VEGETATIVE CHARACTERISTICS: White heath aster's numerous erect, hairy stems sprout from a perennial crown to form a hemispherical clump. The lower leaves, which are about 4 inches (10 cm) long and lance shaped with smooth margins, are typically shed before flowering begins in the fall. The smaller, heathlike leaves on the upper portions of the stem are less than 0.5 inch (1.5 cm) long and persist through the flowering period.

FLOWERS AND FRUIT: From late August or early September through October, white heath aster is covered with masses of tiny, white (occasionally pink), daisylike flower heads in a showy dome-shaped display. The flower heads are about 0.5 inch (1.5 cm) wide and are composed of numerous "petals" (ray florets) with yellow centers (disk florets) that turn reddish following pollination by a variety of insects. The tiny seeds have numerous bristles at the apex that facilitate wind dispersal.

GERMINATION AND REGENERATION: The seeds germinate on bare soil in sunny locations; established plants sprout from a perennial woody crown.

HABITAT PREFERENCES: White heath aster grows best in full sun in dry, sandy soil but can tolerate compacted soils. It is common in unmanaged urban meadows, vacant lots, rubble dump sites, small pavement openings and cracks, and unmowed highway banks.

ECOLOGICAL FUNCTIONS: Tolerant of roadway salt and compacted soil; erosion control on slopes; food and habitat for wildlife.

CULTURAL SIGNIFICANCE: White heath aster is a remarkably tough native species that is strikingly beautiful in flower. It tolerates urban conditions and high soil pH and blooms very late in the season—features that make it a good candidate for cultivation in low-maintenance meadows and landscapes. *Aster* means "star" in Latin.

RELATED SPECIES: Heart-leaved aster (*Symphyotrichum (Aster) cordifolium* (L.) **Nesom**) is a native species that produces pale blue–purple flower heads, about 0.5 inch (1.5 cm) wide, at the ends of the branches from August through October. The yellow disk flowers become reddish following insect-pollination. The plant grows up to 3 feet (1 m) tall and has heart-shaped leaves with conspicuously toothed margins; it is remarkably drought tolerant and flourishes in semishade at the edges of roadways, woodlands, and landscape plantings. **White wood aster (*Eurybia (Aster) divaricata* (L.) Nesom)**, another native species, produces flat clusters of white flower heads from August through October. It grows to about 3 feet (1 m) tall; has heart-shaped, coarsely toothed leaves; and is common in semishade along the edges of disturbed woodlands.

White heath aster in an urban meadow in September

White heath aster foliage in late spring

White heath aster growing on a capped landfill in the middle of Boston Harbor

White heath aster flower heads

Heart-leaved aster foliage

Heart-leaved aster blooming in September from inside a privet hedge

Tanacetum vulgare L.   Common Tansy

SYNONYMS: *Chrysanthemum uliginosum*, golden buttons, buttons

LIFE FORM: **herbaceous perennial**; up to 4 feet (1.3 m) tall

PLACE OF ORIGIN: Europe

VEGETATIVE CHARACTERISTICS: Tansy is an upright plant with smooth, 5-angled stems. The alternate, pinnately compound leaves are 4–8 inches (10–20 cm) long and half as wide; the margins of the leaflets are deeply toothed, producing a fern-like appearance. The bruised foliage is highly aromatic and smells something like camphor.

FLOWERS AND FRUIT: Tansy produces conspicuous, golden yellow inflorescences in a terminal, flat-topped cluster from July through September. The individual flower heads are button shaped, about 0.5 inch (12 mm) wide, and consist entirely of insect-pollinated disk florets. The small seeds that follow lack any obvious dispersal mechanism such as a pappus or barbs.

GERMINATION AND REGENERATION: Tansy seeds germinate in a variety of habitats; established plants produce short rhizomes that develop into extensive, long-lived clumps.

HABITAT PREFERENCES: Tansy is a grassland or meadow species in its native Europe, where it is tolerant of compacted soil, high pH, and full sun. In the urban environment it is common in minimally maintained public parks and open spaces, vacant lots, rubble dumps, urban meadows, and unmowed highway banks.

ECOLOGICAL FUNCTIONS: Disturbance-adapted colonizer of bare ground; soil building on degraded land; erosion control on slopes; food and habitat for wildlife.

CULTURAL SIGNIFICANCE: Tea or oil made from tansy leaves has long been used in Europe to treat dyspepsia, flatulence, jaundice, and sore throat, and as a wash for swellings and inflammations. For culinary purposes, the leaves and shoots are added to a variety of puddings and omelets, and are used to make "tansy cakes," which are eaten at Easter. Tansy was cultivated at John Winthrop's Plymouth colony in New England as early as 1631, and William Darlington, writing in 1847, noted that it "was originally introduced as a garden-plant, and generally cultivated for its aromatic bitter properties—which have rendered it a prominent article in the popular *Materia Medica*. It has now escaped from the gardens, and is becoming naturalized—and something of a weed—in many places." Landscape architecture students collect the dried flower heads to use as trees in their models.

Tansy growth habit

Tansy foliage

Close-up of tansy flower heads

Tansy pollinators in action

Spent seed heads of tansy

Taraxacum officinale Weber ex Wiggers   Dandelion

SYNONYMS: *Leontodon taraxacum*, lions-tooth, blow-ball, cankerwort, pissabed

LIFE FORM: herbaceous perennial; leaves up to 8 inches (20 cm) long

PLACE OF ORIGIN: Eurasia

VEGETATIVE CHARACTERISTICS: Dandelion is low-growing plant that produces a basal rosette of long, dark leaves that are 3–10 inches (7.5–25 cm) long and have deeply lobed margins whose tips point back toward the crown of the plant. The stout, fleshy taproot can be difficult to remove. All parts of the plant exude milky sap when broken.

FLOWERS AND FRUIT: Dandelion produces bright yellow flower heads in early spring and, to a lesser extent, in fall. The heads are produced at the ends of smooth, hollow stalks and consist entirely of petal-like ray florets that are 1–2 inches (3–5 cm) wide. Although often visited by insects, most of the seeds develop *apomictically* (i.e., without the benefit of sexual union) and are arranged into the distinctive globe-shaped "puffballs" that children love to disperse in spring.

GERMINATION AND REGENERATION: The wind-dispersed seeds germinate readily under a wide variety of ecological conditions. Because of the absence of functional sexual reproduction, seedlings are genetically identical to the parent that produced them. Dandelion sprouts readily from a perennial crown as well as from pieces of the root left behind following failed attempts at removal.

HABITAT PREFERENCES: Dandelion tolerates a wide range of growing conditions. It is common in lawns, neglected ornamental landscapes, vacant lots, rubble dumps, small pavement openings, rock outcrops, stone walls, highway banks and median strips, drainage ditches, and railroad rights-of-way. In its native habitat dandelion grows on cliffs and in open woodlands.

ECOLOGICAL FUNCTIONS: Disturbance-adapted colonizer of bare ground; food for wildlife.

CULTURAL SIGNIFICANCE: Dandelion was an early arrival in North America. Josselyn's *New-England's Rarities* (1672) lists it under the category: "Of such plants as have sprung up since the English planted and kept cattle in New England." The young leaves are widely collected for salad greens in early spring (they are rich in vitamins and minerals), dandelion wine can be made from the fermented flowers, and in fall the roots can be roasted and used as a coffee substitute. Tea made from the fresh root acts as a diuretic and has long been used in Europe to treat liver, bladder, and kidney ailments. Homeowners trying to grow the perfect lawn despise dandelion, and spend millions of dollars on herbicides to control its spread. This is quite a turnaround when one considers an early visitor's description of New York's Central Park filled with "blessed dandelions in such beautiful profusion as we have never seen elsewhere, making the lawns in places, like green lakes reflecting a heaven sown with stars" (quoted in Pauly, 2007).

SIMILAR SPECIES: Fall dandelion (*Leontodon autumnalis*) blooms throughout the summer. The lobes of its leaves are rounded rather than sharp pointed, and its brownish seed heads do not form a symmetrical "puffball."

Dandelion in early spring

Dandelions are among the first plants to bloom in early spring

Dandelion foliage rosette in fall

Dandelion being visited by flies

Dandelion seed head

Tragopogon pratensis L. Meadow Salsify

SYNONYMS: Showy goat's beard, yellow goat's beard, yellow salsify, wild oyster plant

LIFE FORM: biennial; up to 3 feet (1 m) tall

PLACE OF ORIGIN: Eurasia

VEGETATIVE CHARACTERISTICS: Meadow salsify is a basal rosette of linear, grass-like leaves in its first year. In its second year the plant produces numerous hollow flowering stems that are smooth, round, and somewhat succulent. The dull green leaves are about a foot (30 cm) long and uniformly narrow with bases that clasp and enclose the stem. The fleshy taproot and all other parts of the plant produce milky sap when broken.

FLOWERS AND FRUIT: Meadow salsify produces solitary yellow flower heads, up to 2.5 inches (6 cm) wide, at the ends of tall flowering stalks in late spring and early summer. The flower heads consist entirely of petal-like ray florets subtended by numerous sharp-pointed, green bracts that are equal in length to the petals of the ray florets. The flower heads turn conspicuously toward the sun and typically remain open until midday. Following pollination by insects, the bracts continue to elongate, eventually forming an elongated cone that encloses the developing seeds. The mature seed heads look like those of dandelion but are much larger—up to 3 inches (8 cm) across—and brown rather than white.

GERMINATION AND REGENERATION: The wind-dispersed seeds can travel more than 800 feet (250 m) from their parent and germinate in sunny, disturbed sites.

HABITAT PREFERENCES: Meadow salsify grows best in full sun and dry soil. It is common in abandoned grasslands, urban meadows, vacant lots, rubble dumps, and at the base of rock outcrops and stone walls.

ENVIRONMENTAL FUNCTIONS: Disturbance-adapted colonizer of bare ground; food for wildlife.

CULTURAL SIGNIFICANCE: The white root of meadow salsify is edible when cooked if it is collected before the flowering stalk develops. The young stems and bases of the lower leaves can also be eaten after boiling. Garden salsify or oyster plant, *Tragopogon porrifolius*, has purplish flowers and produces a large white root that can be eaten throughout the winter.

RELATED SPECIES: Western salsify (*Tragopogon dubius* Scop.), a less common European species, resembles meadow salsify, but the segment of stalk just below the flower head is somewhat expanded and the bracts that subtend the flowers are distinctly longer than the petals (*ray flowers*).

Meadow salsify seed heads

Meadow salsify in flower along a roadside

Meadow salsify foliage

Meadow salsify flower heads

Western salsify flower heads

Tussilago farfara L.   Coltsfoot

SYNONYMS: Coughwort, horsehoof, foalswort, bullsfoot, hallfoot

LIFE FORM: herbaceous perennial; up to 1.5 feet (50 cm) tall

PLACE OF ORIGIN: Eurasia

VEGETATIVE CHARACTERISTICS: Coltsfoot is a stemless plant with large, hoof-shaped leaves that appear to emerge directly from the ground. The leaves have a long petiole, angular teeth along their margins, and are 4–6 inches (10–15 cm) wide with a heart-shaped (cordate) base. The upper leaf surface is smooth and dull green; the lower surface is covered with woolly white hairs that are remarkably soft to the touch.

FLOWERS AND FRUIT: Coltsfoot produces solitary, yellow flower heads at the ends of 6–8 inch (15–20 cm) leafless stalks that are covered with reddish scales. The flower heads, which are about an inch (2.5 cm) wide, emerge in early spring before the leaves appear and consist of numerous yellow "petals" (ray florets) surrounding sterile disk florets. The insect-pollinated flower heads close at night and open during the day. The fluffy white seed heads, which resemble those of dandelion, develop quickly and disperse their pappus-tufted seeds as the leaves are expanding.

GERMINATION AND REGENERATION: The wind-dispersed seeds germinate readily in a variety of habitats. Established plants send out rhizomes to form large, dense colonies.

HABITAT PREFERENCES: Coltsfoot is a highly adaptable plant that grows across a wide range of habitats from full sun to dense shade. It prefers moist soils such as those found along the edges of streams and rivers and in roadside drainage ditches, but it can tolerate high soil pH and will grow in nutrient-poor sandy, clay, or saturated soil. Remarkably, it also grows well in gravel substrates and is particularly common along railroad tracks. In its native habitat in the European Alps it colonizes the bare, rocky soil that is exposed as glaciers retreat.

ECOLOGICAL FUNCTIONS: Disturbance-adapted colonizer of bare ground; erosion control on slopes.

CULTURAL SIGNIFICANCE: Dioscorides included coltsfoot in his first-century herbal, *De Materia Medica*. A tea made from the flowers and leaves has long been used in Europe as a cough suppressant and to treat sore throat, asthma, and lung congestion; it has also been used to flavor a cough-suppressing candy. In *American Weeds and Useful Plants* (1859), William Darlington described coltsfoot as "one of those harmless plants which have long been considered as efficacious domestic remedies, and it is even cultivated in old gardens. An infusion of the whole plant is used for coughs and pulmonary complaints." It is also an excellent source of emergency toilet paper in the field.

Coltsfoot foliage and mature seed heads

Coltsfoot flower heads emerge in early spring, before the leaves

Coltsfoot flowering in the Swiss Alps as the snow is melting

Coltsfoot leaves: upper surface (*left*), lower surface (*right*)

Coltsfoot seed head in late spring

Xanthium strumarium L.   Common Cocklebur

SYNONYM: clotbur

LIFE FORM: summer annual; up to 4 feet (1.3 m) tall

PLACE OF ORIGIN: eastern North America, Europe, and Asia

VEGETATIVE CHARACTERISTICS: The upright stems of cocklebur are well branched, hairy, and distinctly mottled with purple spots. The leaf blades, which are about 6 inches (15 cm) long, are rough to the touch and egg or heart shaped with wavy, irregularly toothed margins.

FLOWERS AND FRUIT: Inconspicuous green flowers are produced from July through September. Wind-pollinated male and female flower heads occur separately on the same plant. The fruit is a hard, egg-shaped bur about an inch (2.5 cm) long that contains 2 seeds. The bur is conspicuously covered with short, hooked spines that facilitate animal dispersal; 2 sharp-pointed "beaks" project from its tip.

GERMINATION AND REGENERATION: The seeds germinate in full sun and dry, well-drained soil. The larger of the 2 seeds in the bur germinates the following spring, but the smaller one typically remains dormant for several years—an interesting example of temporal seed dispersal.

HABITAT PREFERENCES: Cocklebur is particularly common along sandy ocean beaches in full sun, but it also grows in a variety of disturbed urban sites including rubble dumps, vacant lots, and drainage ditches.

ECOLOGICAL FUNCTIONS: Disturbance-adapted colonizer of bare ground; tolerant of compacted soil.

CULTURAL SIGNIFICANCE: Native Americans used a tea made from the leaves to treat kidney problems and diarrhea and as a blood tonic; in European herbal medicine the root has been used to treat swollen lymph nodes and rabies.

Cocklebur growth habit

Spotted stems of cocklebur

Cocklebur growing in sandy coastal soil

Cocklebur in flower

Cocklebur male (*above*) and female (*below*) flower heads

Cocklebur fruits in late summer

Impatiens capensis Meerb.   Jewelweed

Synonyms: *Impatiens fulva, I. biflora,* spotted touch-me-not, hummingbird tree, snapweed, lady's-earrings, touch-me-not

Life Form: summer annual; up to 6 feet (2 m) tall

Place of Origin: eastern North America

Vegetative Characteristics: Jewelweed is an extremely fast-growing plant that produces tall stems with prop roots at their base for added stability. The succulent stems are smooth and hollow and exude a mucilaginous fluid when broken; they are distinctly swollen at the nodes where branches are attached, and are either bluish gray or light green depending on how much white "bloom" is on them. The alternate, oval leaves are thin, hairless, and 1–4 inches (2.5–10 cm) long with coarsely toothed margins; they are remarkable for being "unwettable" because microscopic hairs trap a thin layer of air at the leaf surface. Plants die completely with the first frost.

Flowers and Fruit: Jewelweed produces its conspicuous orange-yellow, 3-petaled flowers from June through September. The funnel-shaped blooms are about an inch (2.5 cm) long and have a downward-curving, nectar-filled spur at the base. They are typically pollinated by bees and are commonly visited by hummingbirds. The flowers dangle freely on long stalks, and their petals are mottled with reddish brown spots. They are followed quickly by small, green, spring-loaded capsules—about an inch (3 cm) long—which explode when touched, sending the seeds flying in all directions (the origin of the scientific and common names). Peeling away the green seed coat reveals a turquoise-blue embryo, the plant's "jewel." The plant also produces small, green cleistogamous flowers that never open and are self-pollinated.

Germination and Regeneration: The seeds germinate in early spring and grow rapidly; there is no regeneration except from seed.

Habitat Preferences: Jewelweed grows best in moist soil in either sun or shade. It is particularly common along the margins of freshwater wetlands and ponds, but also grows in the understory of moist, shady woodlands. It typically forms dense stands and is one of the few plants that can compete with garlic mustard (*Alliaria petiolata*) in the understory of second-growth forests.

Ecological Functions: Stream and river bank stabilization; nutrient absorption in wetlands; food for wildlife (deer love it).

Cultural Significance: Josselyn reported in 1672 that Native Americans used a poultice made from leaves "bruised between two stones" mixed with "hogs grease" to treat bruises and aches. The fresh juice of the leaves has been used to reduce the swelling and inflammation caused by poison ivy. The seeds are reported to be toxic.

Jewelweed
illustrated
in Josselyn's
*New-
England's
Rarities*
(1672)

Jewelweed growth habit

Jewelweed flowers in profile

Close-up of jewelweed flower

Jewelweed seedpods
(exploded and unexploded),
green seeds, and turquoise
blue embryos

Stand of jewelweed growing in a sunny wetland

Alliaria petiolata (Bieb.) Cavara & Grande   Garlic Mustard

Synonyms: *Alliaria officinalis, Sisymbrium alliaria,* jack-by-the-hedge, sauce-alone

Life Form: biennial; up to 3 feet (1 m) tall

Place of Origin: Eurasia

Vegetative Characteristics: The first-year plant is a rosette of leaves 1.25–4.5 inches (3–6 cm) long and wide that stays green throughout the winter. The kidney- or heart-shaped leaves are very distinctive, with scalloped edges, sunken veins, and overlapping bases. Second-year plants produce a tall flower stalk with alternately arranged triangular leaves with short petioles. The whole plant smells of garlic when crushed, and the region where the shoot and root come together (the hypocotyl) typically develops a distinctive S-shaped curve.

Flowers and Fruit: Garlic mustard produces small, white, 4-petaled flowers in terminal clusters in early spring. They can either be insect-pollinated or, if that fails, self-pollinated. The linear fruits—1–2 inches (2.5–5 cm) long—that follow contain 10–20 black seeds that are forcibly ejected when the pods split open in late June or July. The entire plant dies in August, leaving behind tall brown stalks topped with shattered seedpods.

Germination and Regeneration: Seeds germinate in the spring or the following fall after undergoing a period of winter chilling. Seedlings can reproduce under a wide array of ecological conditions, but plants growing in moist, sunny sites are more robust and produce more seeds than those growing in dry, shady sites.

Habitat Preferences: Garlic mustard is a disturbance-adapted plant that is most dominant when growing in woodland understories. It is common in a variety of habitats, including shady roadside edges, forest clearings, degraded woodlands, river and stream banks, and landscape planting beds. The plant gains a competitive advantage by producing chemicals that suppress the growth of the mycorrhizal fungi that live on the roots of other plants.

Ecological Function: Disturbance-adapted colonizer of bare ground.

Cultural Significance: Europeans had a long tradition of eating young garlic mustard leaves, particularly in late winter and early spring when few other greens were available; the plant also had some minor uses in folk medicine. Garlic mustard was first reported growing in North America on Long Island in 1868 but did not draw much attention until the early 1980s when it began spreading rapidly throughout the Northeast, no doubt facilitated by the fact that deer do not eat it. Many states now list garlic mustard as an invasive species.

Stand of garlic mustard

Garlic mustard foliage at the end of its first year

Garlic mustard and jewelweed dominate the understory of a Connecticut woodlot

Garlic mustard at the end of its second year

Garlic mustard flowers

Garlic mustard flowers and developing fruits

Barbarea vulgaris R. Br.   Yellow Rocket

SYNONYMS: St. Barbara's cress, wintercress, rocketcress, spring mustard, scurvy grass

LIFE FORM: **winter annual** or **biennial** or **perennial**; up to 3 feet (1 m) tall

PLACE OF ORIGIN: Eurasia

VEGETATIVE CHARACTERISTICS: During the fall, yellow rocket produces a basal rosette of round, glossy leaves—2–8 inches (5–20 cm) long—that stay green through the winter. The following spring it produces smaller lobed leaves with wavy, toothed margins; the terminal lobe on these leaves is substantially larger than the 1–5 pairs of lateral lobes.

FLOWERS AND FRUIT: Tall, terminal spikes of bright yellow, 4-petaled flowers are produced during the spring (April to June) of the second growing season. They can either be insect-pollinated or, if that fails to occur, self-pollinated. A robust plant can produce hundreds of thin fruits, each about an inch (2.5 cm) long, which split into halves and release small oval seeds.

GERMINATION AND REGENERATION: Seeds germinate readily in spring or fall.

HABITAT PREFERENCES: Yellow rocket grows best on nutrient-rich, sandy or loamy soil. In the urban environment it is common in minimally maintained ornamental plantings, vacant lots, rubble dumps, abandoned grasslands, and unmowed highway banks.

ECOLOGICAL FUNCTION: Disturbance-adapted colonizer of bare ground.

CULTURAL SIGNIFICANCE: The young foliage and flowers are good to eat in the spring and taste something like broccoli rabe. The association with Saint Barbara is probably related to the fact that in Europe the greens emerged around the time of her feast day on December 4.

Yellow rocket flowers

Yellow rocket about to bloom in early spring

Yellow rocket in full bloom

Yellow rocket dies after dispersing its seeds

Yellow rocket seedpods

Capsella bursa-pastoris (L.) Medicus   Shepherd's Purse

SYNONYMS: shepherd's bag, pepper plant, shepherd's heart, pick-pocket

LIFE FORM: winter annual; up 1–2 feet (30–60 cm) tall

PLACE OF ORIGIN: Europe

VEGETATIVE CHARACTERISTICS: Shepherd's purse seeds germinate in fall or spring to produce a basal rosette of deeply lobed leaves that are 2–4 inches (5–10 cm) long and have a large lobe at the apex. This rosette gives rise to a tall flowering stalk in spring, summer, or fall, depending on when the seed germinated. Plants vary greatly in size according to the growing conditions.

FLOWERS AND FRUIT: The small, inconspicuous white flowers are produced on a tall leafless stalk in spring and summer; they are either insect- or self-pollinated. The fruit is a heart-shaped, 2-part, flattened pod about 0.3 inches (8 mm) long. The reddish brown seeds can remain viable in the soil for many years. The plant's common name derives from the shape of the seedpods.

GERMINATION AND REGENERATION: The seeds germinate readily in bare soil and disturbed sites.

HABITAT PREFERENCES: Shepherd's purse is widespread throughout the world. In an urban context it is common in minimally maintained public parks, vacant lots, rubble dumps, minimally maintained lawns, small pavement openings and cracks, ornamental planting beds, rock outcrops, and stone walls.

ECOLOGICAL FUNCTION: Disturbance-adapted colonizer of bare soil.

CULTURAL SIGNIFICANCE: Dioscorides included shepherd's purse in his first-century herbal compendium, *De Materia Medica*. Traditional European herbalists used a tea made from shepherd's purse to stop internal bleeding and hemorrhaging. The plant was an early arrival in North America and is listed in Josselyn's *New-England's Rarities* (1672) under the category: "Of such plants as have sprung up since the English planted and kept cattle in New England." The young shoots are edible in spring when cooked like spinach.

Shepherd's purse in silhouette

Shepherd's purse basal rosette

Dense stand of shepherd's purse

Shepherd's
purse growth
habit and
taproot

Shepherd's
purse flowers
and rapidly
developing
fruits

Cardamine hirsuta L.   Hairy Bittercress

SYNONYMS: hoary bittercress, lamb's cress, land cress, shot weed

LIFE FORM: winter or summer annual; up to a foot (30 cm) tall

PLACE OF ORIGIN: Eurasia

VEGETATIVE CHARACTERISTICS: Hairy bittercress seeds typically germinate in autumn and produce a basal rosette of pinnately compound leaves with 2–10 pairs of round to kidney-shaped leaflets. They are 2–4 inches (5–10 cm) long and quite conspicuous relative to the smaller stem leaves.

FLOWERS AND FRUIT: In early spring, hairy bittercress rosettes give rise to smooth flowering stalks with alternate leaves that range in height from 2–12 inches (5–30 cm). Each flower stalk is topped with a dense cluster of tiny white flowers, each with 4 petals. The flowers are either insect- or self-pollinated and quickly produce slender pods about an inch (2.5 cm) long that explosively disperse their seeds at the slightest touch. A single vigorous plant can produce a dozen flower stalks and hundreds of seeds.

GERMINATION AND REGENERATION: Hairy bittercress seeds germinate in fall and spring. The plant completes its life cycle very quickly and under cool, moist conditions is capable of producing multiple generations in a single season.

HABITAT PREFERENCES: Hairy bittercress grows best in damp, disturbed soil in either sun or shade but is also found in dry, sunny sites. It is common in cultivated garden beds where the soil is bare or has been turned over.

ENVIRONMENTAL FUNCTION: Disturbance-adapted colonizer of bare ground.

CULTURAL SIGNIFICANCE: The rosettes can be eaten in spring as a bitter herb. Left to its own devices, the plant will quickly dominate an untended vegetable garden.

Hairy bittercress seedpods

Hairy bittercress in early spring

Hairy bittercress rosette in fall

Hairy bittercress flowers and rapidly developing seedpods

Hairy bittercress seedpods about to explode

Lepidium virginicum L.   Virginia Pepperweed

SYNONYMS: pepper-grass, poor-man's pepper

LIFE FORM: winter or summer annual; up to 2 feet (70 cm) tall

PLACE OF ORIGIN: eastern North America

VEGETATIVE CHARACTERISTICS: In spring Virginia pepperweed is a basal rosette of smooth, bright green leaves that have a large terminal lobe and numerous small lateral lobes. Later in the season the plant produces a smooth, erect flowering stem with small narrow leaves about an inch (2.5 cm) long that do not clasp the stem.

FLOWERS AND FRUIT: Virginia pepperweed produces dense spikes of small white to greenish flowers on long, highly branched stalks (racemes) from June through September; they are either insect- or self-pollinated. These are followed quickly by small, slightly winged fruits with a notch at the apex that contain abundant seed. In seed the plant appears to be covered with stiff bottle-brushes.

GERMINATION AND REGENERATION: The seeds germinate in fall or spring.

HABITAT PREFERENCES: This plant is ubiquitous in sunny, compacted soil, including small pavement openings, neglected residential and commercial landscapes, minimally maintained public parks, vacant lots, rubble dump sites, and unmowed highway banks and median strips.

ECOLOGICAL FUNCTION: Disturbance-adapted colonizer of bare soil.

CULTURAL SIGNIFICANCE: William Darlington reported in 1859, "This common weed is a native of the southern portion of our country, and is abundantly naturalized in many parts of Europe—thus making a partial return for the abundant supply of weeds which has crossed the ocean to our shores." Native Americans used Virginia pepperweed to treat poison ivy rash, scurvy, and the croup. The flattened pods have a strong peppery taste when eaten, as children (and some birds) like to do. The young shoots are edible in spring.

RELATED SPECIES: Field pepperweed (*Lepidium campestre* (L.) R. Br.) is a European winter annual with densely hairy, gray-green foliage and stiff, branched stems that grow to about 2 feet (60 cm) tall; the lower leaves are lance shaped, up to 2 inches (5 cm) long, and taper to the base; the upper stem leaves are distinctly alternate and clasp the stem. The white flowers are produced on erect spikes in early summer and are followed by "bottle-brush" clusters of distinctively winged seedpods. Field pennycress (*Thlaspi arvense* L.), another European winter annual, is about 2 feet (60 cm) tall and has coarsely toothed, lance-shaped leaves that clasp the stem at their base. It produces dense racemes of small white flowers at the top of the stems from early spring through early summer; these are followed by distinctive flat, rounded pods about 0.5 inch (1.3 cm) in diameter with a notch at the apex. The young leaves are edible and taste something like mustard.

Virginia pepperweed with developing seedpods

Virginia pepperweed loaded with seed in its typical urban habitat

Virginia pepperweed in flower

Field pepperweed in seed; note the clasping leaves

Field pennycress in a concrete drainage pipe planter

Field pennycress with flowers and developing seedpods

Cerastium fontanum (Hartman) Gretyker & Byrdet   Mouseear Chickweed

Synonyms: *Cerastium vulgatum*, mouseear, chickweed

Life Form: **herbaceous perennial**; up to 1 foot (30 cm) tall

Place of Origin: Eurasia

Vegetative Characteristics: Mouseear chickweed produces distinctly hairy leaves that are about half an inch (12 mm) long and an eighth of an inch (6 mm) wide; they have an opposite arrangement, lack a distinct petiole (sessile) and are oblong to oval. The hairy stems, which are rough to the touch, are initially upright but with time (and repeated mowing) become horizontal, forming mats up to 2 feet (70 cm) wide. Mouseear chickweed typically remains green through the winter.

Flowers and Fruit: White flowers about half an inch (12 mm) wide are produced from May through October in loose clusters at the ends of the branches. The blossoms have 5 petals, each of which is deeply notched at the tip, making it look like there are twice that number; they can either be insect- or self-pollinated. The fruit is a cylindrical capsule full of tiny reddish brown seeds.

Germination and Regeneration: Seedlings germinate throughout the year in a variety of habitats. Established plants form perennial crowns that produce new shoots that can root wherever they touch the ground, facilitating the plant's spread across the landscape.

Habitat Preferences: Mouseear chickweed grows best in sunny, moist soil but also tolerates dry shade. It is common in minimally maintained lawns (where it can survive mowing and form large patches); pavement openings and sidewalk cracks; vacant lots and dumps; and roadsides and median strips, where its tolerance of road salt is evident.

Environmental Function: Disturbance-adapted colonizer of bare ground.

Developing fruits of mouseear chickweed

Close-up of mouseear chickweed flowers

Mouseear chickweed foliage

Close-up of mouseear chickweed leaves

Mouseear chickweed flowering among tall grasses by the side of a highway

Saponaria officinalis L.   Bouncing Bet

SYNONYMS: soapwort, old maid's pink, hedge pink, bruisewort, fuller's herb

LIFE FORM: **herbaceous perennial**; up to 2 feet (70 cm) tall

PLACE OF ORIGIN: Eurasia

VEGETATIVE CHARACTERISTICS: Bouncing bet typically appears as a cluster of unbranched stems. The leaves are smooth, opposite, dark green, 1.5–3 inches (3–6 cm) long, and oval to lance shaped. The leaf base forms a collar around the stem, giving it a somewhat jointed appearance.

FLOWERS AND FRUIT: Bouncing bet produces terminal clusters of pale pink to cream-colored flowers from July through September. Individual flowers are very showy, about an inch (2.5 cm) across, and have notched petals that are often bent back toward the stem (reflexed). The flowers are sweetly fragrant at the end of the day and are probably pollinated by night-flying moths. Mutant individuals with double flowers (i.e., multiple petals) commonly grow alongside plants with normal flowers.

GERMINATION AND REGENERATION: The seeds germinate readily in full sun; established plants produce new stems from short rhizomes, forming large clumps.

HABITAT PREFERENCES: Bouncing bet grows best in dry, sandy or gravelly soil and full sun. It is common along railroad tracks and in vacant lots, rubble dumps, urban meadows, roadside drainage ditches, and neglected ornamental landscapes.

ECOLOGICAL FUNCTION: Disturbance-adapted colonizer of bare ground.

CULTURAL SIGNIFICANCE: One of the common names of this plant, soapwort, comes from the fact the root, when bruised and moistened with warm water, produces a lather that can be use for washing and cleaning. The active ingredient belongs to a class of compounds known as *glycoside saponins*, which are toxic to both animals and people. The ancient Romans used the plant for washing and shrinking wool, and the Greek physician Dioscorides included it in first-century herbal *De Materia Medica*. European herbalists used bouncing bet externally to treat skin rashes, most notably "the itch" caused by syphilis and psoriasis; because of its toxicity it had only limited internal use. Bouncing bet was intentionally introduced into North America at an early date because of its detergent and medicinal attributes. A double-flowered cultivar ('Flore Pleno') is available in the nursery trade.

Bouncing bet grows well in sandy soils like those found on Cape Cod, Massachusetts

Bouncing bet foliage

Normal flowers of bouncing bet

Bouncing bet commonly produces double-flowered individuals

Bouncing bet growing along the roadside

Scleranthus annuus L.   Knawel

SYNONYM: German knotgrass

LIFE FORM: **winter annual**; up to 4 inches (10 cm) tall

PLACE OF ORIGIN: Eurasia

VEGETATIVE CHARACTERISTICS: Knawel (pronounced *nôl*) is a mat-forming plant with short, linear leaves that are opposite and range in length from one-eighth of an inch to three-quarters of an inch (5–20 mm); lateral branches arise spontaneously from the leaf axils. Under the proper conditions knawel can form spreading clumps more than a foot (30 cm) wide with a distinctive gray-green color; it produces a small, tenacious taproot.

FLOWERS AND FRUIT: Knawel produces tiny, inconspicuous green flowers from May through October. They are difficult to see because they are sessile in the leaf axils and lack petals. The flowers are mainly self-pollinated and are quickly followed by tiny urn-shaped fruits, each containing a single seed.

GERMINATION AND REGENERATION: Seeds germinate readily in sunny, dry locations.

HABITAT PREFERENCES: Knawel is ubiquitous along roadsides and sidewalks throughout the urban environment. It flourishes in dry soil and full sun and is highly tolerant of road salt. It is commonly found in the compacted soil adjacent to paved areas. The common name comes from the German word for a knot or ball of yarn.

ENVIRONMENTAL FUNCTION: Disturbance-adapted colonizer of bare ground.

RELATED SPECIES: **Birdseye pearlwort (*Sagina procumbens* L.)** is a tiny, evergreen perennial, seldom more than an inch (2.5 cm) tall, that grows throughout the urban environment in pavement cracks, median strips, and between walkway pavers. Its leaves are about 0.5 inch (1.2 cm) long and are arranged in a whorl, giving the plant a star-shaped appearance. The flowers are inconspicuous, with 4 green sepals that are about one-tenth of an inch (2.5 mm) long. Birdseye pearlwort reproduces by both seeds and stolons, and is tolerant of roadway salt and compacted soil. In its native Eurasian habitat it is a cliff-dwelling species.

Knawel growing
by a roadside

Knawel grows where nothing else can

Tiny flowers of knawel

Birdseye
pearlwort
grows well
between
walkway
bricks

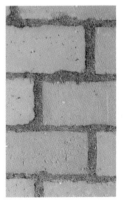

Close-up of birdseye pearlwort foliage

Silene latifolia Poir.   White Campion

SYNONYMS: *Lychnis alba, Silene alba,* white cockle

LIFE FORM: **biennial** or **short-lived perennial**; up to 3 feet (1 m) tall

PLACE OF ORIGIN: Europe

VEGETATIVE CHARACTERISTICS: White campion initially produces a basal rosette of dull green, oval to lance-shaped leaves that are up to 5 inches (10 cm) long and soft-hairy to the touch. Later in the season it sends up a branched stem with opposite leaves and terminal flowers. The roots are thick and fleshy.

FLOWERS AND FRUIT: Loose clusters of white flowers 0.75–1.5 inches (2–4 cm) wide are produced on the end of tall stalks from May through early fall. Their most conspicuous feature is the fused sepals (calyx), which form a prominently ribbed, hairy "bladder" out of which emerge the 5 deeply notched petals. Sweet-scented male and female flowers are produced on separate plants (dioecious) and open in the evening for insect-pollination. At maturity, the light brown, urn-shaped capsules release their small brown seeds through an opening at the top that is surrounded by a crown of 10 teeth.

GERMINATION AND REGENERATION: Seeds germinate readily; established plants disturbed by cultivation can sprout from stem or root fragments.

HABITAT PREFERENCES: White campion grows best in full sun in rich, well-drained soil. In the urban environment it is common along roadsides, in waste places, and in disturbed meadows.

ECOLOGICAL FUNCTION: Disturbance-adapted colonizer of bare ground.

RELATED SPECIES: **Bladder campion** or **maiden's tears (*Silene vulgaris* (Moench) Garcke)** is a rhizomatous perennial with smooth leaves and gray-green stems. In midsummer it produces a profusion of white, insect-pollinated flowers that emerge from inflated, purplish or greenish bladders. The flowers are smaller than those of white campion, but the bladders are hairless and more rounded than elongated. Bladder campion grows best in gravelly soils in full sun and is common in vacant lots and urban meadows. The plant apparently can accumulate zinc when growing in soils contaminated with that element. The young shoots can be eaten in spring, but they are somewhat bitter.

White campion rosette foliage

Close-up of white campion flower showing inflated calyx

Bladder campion growth habit

White campion seedpod, about an inch (2.5 cm) long

Bladder campion in its typical urban habitat

Bladder campion flowers

Spergularia rubra (L.) J. & C. Presl.   Red Sandspurry

SYNONYMS: *Arenaria rubra,* roadside sandspurry, common sandspurry

LIFE FORM: **winter annual** or **short-lived perennial**; up to 6 inches (15 cm) tall

PLACE OF ORIGIN: Europe

VEGETATIVE CHARACTERISTICS: Red sandspurry may be either prostrate or upright depending on the growing conditions. The fine, narrow leaves—about 0.5 inch (12. mm) long—are pointed and flat with an opposite or, more typically, whorled arrangement.

FLOWERS AND FRUIT: Red sandspurry produces pinkish red flowers in May and June and again in late summer and fall; they have five 5 petals alternating with longer green sepals and are 0.5 inch (1 cm) or less in diameter; they can be either insect- or self-pollinated. The plant attracts attention only when it is in flower. The fruit is a small capsule about 0.25 inch (5 mm) long that sticks up above the foliage and is filled with numerous tiny brown seeds.

GERMINATION AND REGENERATION: Seeds germinate readily in a variety of disturbed conditions.

HABITAT PREFERENCES: Red sandspurry grows best in full sun and moist or dry soil. Because of its salt tolerance it is common in roadside drainage ditches, compacted lawns, median strips, small pavement openings, and at the base of rock outcrops and stone walls. This inconspicuous plant grows all over the urban environment but is seldom noticed. In its native European habitat red sandspurry grows along river and stream banks in gravelly or sandy soils.

ECOLOGICAL FUNCTION: Disturbance-adapted colonizer of bare ground.

CULTURAL SIGNIFICANCE: Red sandspurry has been used in traditional European medicine to treat bladder problems.

Growth habit of red sandspurry (about 4 inches [10 cm] tall)

Red sandspurry growing at the Vince Lombardi Rest Area along the New Jersey Turnpike

Red sandspurry flowers

Red sandspurry growing in compacted soil

Close-up of red sandspurry flowers

Stellaria media (L.) Vill.   Common Chickweed

Synonyms: *Alsine media*, starwort, starweed, winterweed, tongue-grass, chicken weed, skirt buttons, chickwhirtles

Life Form: winter annual; up to 1.5 feet (45 cm) tall

Place of Origin: Eurasia

Vegetative Characteristics: Under adverse growing conditions chickweed is a prostrate, mat-forming plant with densely branched stems; in the absence of environmental stress the plant is lush and tall. The opposite leaves are broadly elliptical to egg shaped, an inch or so (2–3 cm) long, and pointed at the tip. All parts of the plant are bright green and smooth except for a fine line of hairs that runs up one side of the stem only and then switches to the opposite side when it reaches a pair of leaves. The root system is shallow and weak. Chickweed grows best under cool, moist conditions in spring, early summer, and fall.

Flowers and Fruit: Common chickweed produces its small, white flowers at virtually any time of the year, depending on the location; in the Northeast this is from February through November. They are about 0.25 inch (4–6 mm) in diameter and have 5 white petals that are shorter than the sepals and so deeply notched that it looks like there are 10 of them. The scientific name of the plant, *Stellaria*, is derived from the Latin word for "star," which describes the shape of the flowers and which mainly self-pollinate. The fruit is a 1-celled capsule containing numerous tiny seeds.

Germination and Regeneration: Seeds germinate readily in cool weather in either the fall or spring; buried seeds can remain viable for many years. Plants typically die in the heat and drought of summer. Because its stems root at the nodes, chickweed can tolerate some mowing.

Habitat Preferences: Common chickweed grows best in moist, nutrient-rich sites but is by no means limited to such areas. It is common in minimally maintained lawns (where it can form large patches), vacant lots, rubble dumps, the margins of freshwater wetlands and streams, stone walls, rock outcrops, small pavement openings, and sidewalk cracks.

Ecological Functions: Disturbance-adapted colonizer of bare ground; food for wildlife, especially birds.

Cultural Significance: Chickweed is a truly cosmopolitan plant that can be found from high elevations in the tropics to high latitudes in Alaska and Eurasia. It is common across North America and is capable of blooming at any time of year. The young leaves and stems are eaten in rural areas throughout the world and taste like baby spinach. Chickweed has long been used as a folk medicine in Europe (Dioscorides described it in detail), Asia, and North America as an ointment for treating indolent ulcers, skin rashes, insect bites, and inflammation. Chickweed was an early arrival in North America and is listed in Josselyn's *New-England's Rarities* (1672) under the category: "Of such plants as have sprung up since the English planted and kept cattle in New England."

Chickweed in a typical urban niche

Chickweed growth habit

Chickweed in early spring

Chickweed foliage

Close-up of chickweed flowers

Chenopodium album L.   Common Lambsquarters

SYNONYMS: fat hen, pigweed, mealweed, goosefoot, bacon-weed, wild spinach

LIFE FORM: **summer annual**; up to 6 feet (2 m) tall

PLACE OF ORIGIN: Europe

VEGETATIVE CHARACTERISTICS: The stems have conspicuous grooves and are usually green, but sometimes show some purple coloration at the point where the branches are attached. The plant branches freely and develops a broad, pyramidal shape at maturity. The alternate leaves are dull green, 2–4 inches (5–10 cm) long, roughly triangular to rhomboidal (hence the common name goosefoot), and have irregular teeth and gray to white, "mealy" undersides. The whole plant has a semi-succulent appearance and sometimes turns purplish at the end of the growing season. Lambsquarters produces a short, tenacious taproot.

FLOWERS AND FRUIT: Large, pale green inflorescences terminate the branches in late summer; the bisexual flowers lack petals, are inconspicuous, and are mainly self-pollinated. The fruits are tiny, bladderlike structures containing a single seed. A large plant can produce up to 75,000 seeds.

GERMINATION AND REGENERATION: Seeds fall to the ground at maturity and germinate in early summer; they are also eaten and dispersed by ground-feeding birds. Buried seeds can remain viable in the soil for decades if not centuries.

HABITAT PREFERENCES: Lambsquarters tolerates a wide variety of soil types and moisture and light regimens, but reaches its full potential in rich soil. The plant is noteworthy for its ability to remain green after other plants have "browned out" from drought or frost. It is common in all sorts of disturbed sites, including neglected ornamental landscapes, minimally maintained public parks, vacant lots, rubble dumps, small pavement openings, chain-link fence lines, rock outcrops, stone walls, unmowed highway banks and median strips, and railroad rights-of-way.

ECOLOGICAL FUNCTIONS: Tolerant of compacted soil; food and habitat for wildlife; soil building on degraded land; phytoremediation in degraded urban landscapes by absorbing heavy metals (zinc, copper, and lead) and binding them to organic matter.

CULTURAL SIGNIFICANCE: Young lambsquarters shoots are edible in the spring after the fine powder that typically covers the leaves is washed away. In times of famine in Europe the seeds were boiled to make gruel or baked into bread (Napoleon and his troops had to live on this at times). The grain quinoa, which has recently become popular as a health food, is the seed of *Chenopodium quinoa*, a species cultivated at high elevations by the Incas.

RELATED SPECIES: **Mexican tea** or **wormseed** (*Chenopodium ambrosioides* L.) is an upright plant that can grow up to 3 feet (1 m) tall and is native to Central and South America. It produces small, bright green, narrow leaves with wavy margins; while it grows best in full sun, it tolerates shade. The foliage emits a pungent camphor or anise-like odor when crushed, and the seeds have long been used to expel worms, especially in children. Mexicans call the plant *epazote* and use it as an aide to digestion and typically add it to chili sauces and bean dishes to reduce flatulence.

Lambsquarters in full flower

Lambsquarters flowers

Lambsquarters growing between a
sidewalk and curb

Lambsquarters foliage

Lambsquarters seeds are included free of
charge in most topsoil

Mexican tea growth habit

Hypericum perforatum L.　Common St. Johnswort

SYNONYMS: Klamath weed, goatweed, rosin weed, tipton weed, god's wonder plant, devil's scourge

LIFE FORM: **herbaceous perennial**; up to 3 feet (1 m) tall

PLACE OF ORIGIN: Europe

VEGETATIVE CHARACTERISTICS: The smooth, upright stems of St. Johnswort are highly branched and somewhat woody at the base. The paired opposite leaves lack a distinct petiole and are arranged perpendicular to each other. They are elliptical to oblong, 1–1.5 inches (2–4 cm) long by 0.25–0.5 inch (5–10 mm) wide, and are covered with numerous tiny translucent dots that are visible when the leaf is held up to the light (hence the Latin name *perforatum*).

FLOWERS AND FRUIT: The flat-topped terminal clusters of bright yellow, insect-pollinated flowers are produced from June through September. The individual flowers are 0.75–1 inch (2–2.5 cm) wide, and the margins of the 5 petals are marked with numerous tiny black spots. Rubbing the flower petals between the fingers produces a reddish stain. The sharp-pointed fruit capsules contain hundreds of tiny seeds that stick to animal fur and human clothing.

GERMINATION AND REGENERATION: Seeds germinate readily but are also capable of germinating after long periods of burial. Established plants produce new shoots from rhizomes that emerge from the perennial crown.

HABITAT PREFERENCES: St. Johnswort grows best in rich soil in full sun but also tolerates dry, sandy or compacted soils. In the urban environment it is common in neglected residential and commercial landscapes, minimally maintained public parks, vacant lots, rubble dumps, abandoned grasslands, meadows, rock outcrops, stone walls, unmowed highway banks and median strips, and railroad rights-of-way.

ECOLOGICAL FUNCTIONS: Disturbance-adapted colonizer of bare ground; food for pollinating and herbivorous insects; tolerant of roadway salt and compacted soil.

CULTURAL SIGNIFICANCE: St. Johnswort has an ancient history of medicinal use in Europe. It appears in Dioscorides' first-century herbal, *De Materia Medica*, and was traditionally used as a poultice for treating wounds and ulcers in European folk medicine. More recently it has become popular as an antidepressant and sedative, marketed as "nature's Prozac." It has also been used as a food preservative, especially for cheese, and as a source of reddish dye. It arrived early in North America and is listed in Josselyn's *New-England's Rarities* (1672). St. Johnswort has never been popular with farmers, mainly because it is poisonous to livestock. John Bartram said of it in 1759: "The common English hipericum is a very pernicious weed. It spreads over whole fields & spoils their pasturage not only by choaking ye grass but infecting our horses & sheep with scaped noses & feet especially those that have white hair on their face & legs."

St. Johnswort
growing in
compacted soil

St. Johnswort flowers

St. Johnswort
in full bloom

Developing
fruits of St.
Johnswort

St. Johnswort in August

Calystegia sepium (L.) R. Br.   Hedge Bindweed

Synonyms: *Convolvulus sepium*, wild morning glory, devil's vine, giant bindweed, lady's nightcap

Life Form: herbaceous perennial vine; stems up to 10 feet (3 m) long

Place of Origin: eastern North America and Europe

Vegetative Characteristics: Hedge bindweed produces smooth stems that trail along the ground, twining in a counterclockwise direction until they encounter something to climb on. Once it reaches the top of a plant or a fence, hedge bindweed engulfs its host with a cascade of leafy stems. The leaves are smooth, alternate, 2–4 inches (5–10 cm) long and less than half as wide, and have a sharp-pointed tip and squared-off basal lobes.

Flowers and Fruit: Hedge bindweed produces solitary, morning glory–like flowers from July through August in the leaf axils; the petals are white to pink and fused into an upward-facing, funnel-shaped tube about 1.5–3 inches (3–6 cm) long and 2 inches (5 cm) wide. The scentless flowers are open early in the day and are pollinated by bumblebees, honeybees, and hawkmoths. The 2 prominent leafy bracts that clasp the base are a distinctive feature of the flower. The fruit is an egg-shaped capsule containing 2–4 dark seeds that fall to the ground at maturity.

Germination and Regeneration: The seeds germinate readily in a variety of soils and disturbed habitats; established plants produce long, fleshy rhizomes that give rise to new shoots.

Habitat Preferences: Hedge bindweed grows best in moist soils but is not limited to them. In the urban environment it is common along the margins of wetlands and streams, stone walls, rock outcrops, ornamental planting beds, roadsides, and chain-link fences.

Ecological Functions: Tolerant of roadway salt and compacted soil; erosion control.

Cultural Significance: Dioscorides included hedge bindweed in *De Materia Medica*, which was written in the first century AD. The root has been used in traditional medicine as a purgative and to treat jaundice and gall bladder problems.

Related Species: Field bindweed (*Convolvulus arvensis* L.), a perennial, herbaceous vine from Eurasia, is less of a climber than hedge bindweed and commonly forms a ground-covering mat of vegetation. It produces numerous small white or pink flowers about 0.75 inch (2 cm) long, roughly half the size of those produced by hedge bindweed. The leaves are about 1.5–2.5 inches (4–6 cm) long with pointed lobes at the base and resemble arrowheads. The plant reproduces vigorously from deep-growing roots and is a serious agricultural weed in many parts of the world. **Tall morning glory (*Ipomoea purpurea* (L.) Roth)**, an annual vine native to Central America, has alternate, smooth, heart-shaped leaves and purple to pale blue funnel-shaped flowers that are 1–2 inches (2.5–5 cm) long and wide. It prefers moist, rich soil but grows in a variety of habitats; it commonly climbs on chain-link fences and is cultivated by many homeowners.

Hedge bindweed foliage

Hedge bindweed flowers

Hedge bindweed taking over an arborvitae hedge

Field bindweed growth habit

Field bindweed flowers

Tall morning glory leaves and flowers

Echinocystis lobata (Michx.) T. & G.   Wild Cucumber

SYNONYMS: wild balsam apple, squirting cucumber

LIFE FORM: **summer annual vine**; stems more than 10 feet (3 m) long

PLACE OF ORIGIN: eastern North America

VEGETATIVE CHARACTERISTICS: Wild cucumber produces smooth, thin, green stems; the leaves are alternate, smooth, star shaped—mostly with 5 sharp-pointed lobes—and about 6 inches (15 cm) long. The plant climbs by means of forked tendrils that coil tightly once they find an object to attach to. The root system is remarkably spindly given the extent of the aboveground growth. The whole plant dies with the first frost.

FLOWERS AND FRUIT: Wild cucumber is monoecious, producing separate male and female flowers on the same plant in August. The flowers of both sexes are yellowish white and have 6 petals; the showy male flowers are clustered together on erect spikes about 6–8 inches (15–20 cm) long; the inconspicuous female flowers are produced on short spikes. Insects, most likely bees, are responsible for pollination. The fruit is a spiny green bladder about 2 inches (5 cm) long by 1 inch (2.5 cm) wide.

GERMINATION AND REGENERATION: Four flat, pumpkin-like seeds are forcibly ejected from the mature fruit as it dries out, leaving behind a hollow, skeletonized shell that resembles a miniature loofah sponge dangling from the dry stalk. Seeds germinate in late spring in the shade of other plants and rapidly climb up their stems to reach the light.

HABITAT PREFERENCES: Wild cucumber grows best on moist ground in thickets with partial shade; it wilts noticeably in full sun, even when there is adequate soil moisture. In the urban environment it is found on stockpiled topsoil and compost piles, climbing over other plants along the margins of streams and wetlands, and scrambling through and covering over ornamental plantings.

ECOLOGICAL FUNCTION: Disturbance-adapted colonizer of moist, shady ground.

CULTURAL SIGNIFICANCE: Native Americans made a bitter tea from the root to treat stomach troubles, kidney ailments, rheumatism, and as a love potion. Charles Darwin studied and wrote about the climbing habits of this plant.

RELATED SPECIES: Burcucumber (*Sicyos angulatus* L.), another summer annual, tendril-climbing vine, is native to eastern North America. Its leaves are alternate and broadly 5-lobed with heart-shaped bases and an overall pentagon shape, as opposed to the star-shaped leaves of wild cucumber. Female and male flowers are produced separately on the same plant; both are greenish white and not nearly as showy as those of wild cucumber. The fruits are produced in clusters on a single stalk and look like small, prickly cucumbers about 0.5 inch (1–2 cm) long. The plant is common along the sunny margins of streams, wetlands, and woodland thickets.

Wild cucumber smothering the surrounding vegetation

Mature wild
cucumber fruit

Wild cucumber with male flowers scrambling over a
clump of scouring rush

Burcucumber on a stone wall in
Boston

Burcucumber growth habit

Burcucumber fruits

Acalypha rhomboidea Raf.   Rhombic Copperleaf

Synonyms: three-seeded mercury, mercury weed

Life Form: summer annual; up to 2 feet (70 cm) tall

Place of Origin: eastern North America

Vegetative Characteristics: Rhombic copperleaf produces hairy to partly hairy stems. The broadly lance- or diamond-shaped (rhombic) leaves are 1–3 inches (2–8 cm) long, alternately arranged, and have marginal teeth and petioles up to 1.5 inches (4 cm) long. The young leaves often develop a distinct copper color, and the entire plant turns coppery as it begins to die in the fall. Typically, plants develop a single taproot with weak lateral roots.

Flowers and Fruit: Rhombic copperleaf produces separate male and female flowers on the same plant (monoecious) from June through October. The flowers are green and are located in the axils of the uppermost leaves; they are inconspicuous at the time of pollination (probably by wind) and are surrounded by a deeply notched bract with 5–9 lobes. As the female flowers develop into fruits the bracts become quite prominent and leaflike, often taking on a coppery color as they reach maturity.

Germination and Regeneration: Seeds germinate in late spring when soil temperatures start to warm up.

Habitat Preferences: Rhombic copperleaf grows well under a wide variety of moisture and nutrient conditions in full sun or partial shade. It is common along the unmaintained edges of sidewalks, in roadways and horticultural plantings, along shady stream banks, and in the understory of open woodlands.

Ecological Function: Disturbance-adapted colonizer of bare ground.

Related Species: Virginia copperleaf (*Acalypha virginica* L.) is very similar to rhombic copperleaf, and the two are often confused. The bracts surrounding the flowers and fruits of Virginia copperleaf have more lobes (9–15) than those of rhombic copperleaf, and the leaf petioles are typically less than half as long as the leaf blades. **Slender copperleaf** (*Acalypha gracilens* A. Gray) produces narrow leaves about 2 inches (5 cm) long by 0.5 inch (1 cm) wide, with petioles that are short relative to the length of the blade. All three species of *Acalypha* are native to eastern North America.

Rhombic copperleaf in the urban environment

Rhombic copperleaf foliage

Rhombic copperleaf growing as a weed in
a pot of plastic flowers

Leafy bracts and flowers of rhombic
copperleaf

Rhombic copperleaf
(*left*) and slender
copperleaf (*right*)

Chamaesyce maculata (L.) Small   Spotted Spurge

SYNONYMS: *Euphorbia maculata*, *Euphorbia supina*, prostrate spurge, creeping spurge, matweed, milk purslane, spotted sandmat

LIFE FORM: **summer annual**; several inches (5–10 cm) tall and up to 16 inches (40 cm) wide

PLACE OF ORIGIN: eastern North America

VEGETATIVE CHARACTERISTICS: Spotted spurge appears late in the spring and disappears with the first frost. It grows close to the ground, forming a dense mat with branches arranged in an almost perfect circle. The plant is extremely variable in the morphology of its tiny leaves, which are opposite, oblong to linear, and one-eighth to five-eighths inch (5–15 mm) long and half as wide. The leaves typically have a distinctive maroon blotch on the upper surface and reddish purple undersides, although the leaves of some plants are reddish purple on both the upper and lower surfaces, and some lack the distinctive maroon blotch altogether. Spotted spurge stems are pink to red, densely hairy, and exude milky sap when broken. The plant produces a weak taproot, and the prostrate stems do not produce adventitious roots. Although the leaves are opposite, the branching usually appears alternate because only one bud from each leaf pair typically produces a branch.

FLOWERS AND FRUIT: The inconspicuous greenish flowers are produced from July to September in the leaf axils; they lack petals and sepals and are either self- or insect-pollinated. The fruit is a tiny 3-seeded capsule.

GERMINATION AND REGENERATION: The seeds germinate readily in full sun and dry, low-nutrient soils.

HABITAT PREFERENCES: Spotted spurge grows best in disturbed sites in full sun. It tolerates both drought and soil compaction and is common in scale pavement openings and cracks, vacant lots, rubble dumps, trampled lawns, neglected ornamental landscapes, vegetable gardens, rock outcrops, and stone walls.

ECOLOGICAL FUNCTION: Disturbance-adapted colonizer of bare ground.

CULTURAL SIGNIFICANCE: Native Americans used spotted spurge as a blood purifier and to treat urinary problems. The plant is toxic to livestock, and its milky sap can irritate the skin in people. Spotted spurge grows throughout most of North America.

SIMILAR SPECIES: Carpetweed, purslane, and prostrate knotweed are other examples of mat-forming plants with a circular growth pattern and a distinct taproot that seem "preadapted" to grow in sidewalk cracks and survive foot traffic. Of the four, only spotted spurge has milky sap.

Spotted spurge growing between a sidewalk and curb

Spotted spurge in a parking lot crack

Close-up of spotted spurge leaves and flowers

Spotted spurge foliage and developing fruits

Spotted spurge displays distinct alternate branching

Euphorbia cyparissias L.   Cypress Spurge

Synonym: graveyard spurge

Life Form: herbaceous perennial; up to a foot (30 cm) tall

Place of Origin: Eurasia

Vegetative Characteristics: The short, narrow, gray-green leaves are about an inch (2.5 cm) long and are packed tightly around the stem, giving them a distinct bottle-brush appearance. The stems and leaves exude milky sap when broken, a characteristic of the spurge family.

Flowers and Fruit: Cypress spurge produces tiny clusters of insect-pollinated flowers subtended by conspicuous, bright yellow (occasionally pink), petal-like bracts from late spring through summer. The fruits are tiny capsules, each containing 3 brown seeds.

Germination and Regeneration: Seeds germinate in sunny, dry sites and are not typically produced in cold climates such as New England. The plant reproduces strongly from root suckers and can be extremely difficult to eradicate.

Habitat Preferences: Cypress spurge grows best in sunny, dry, sandy or rocky soils. In the urban environment it is common along railroad tracks, rock outcrops, stone walls, and in small pavement openings.

Ecological Functions: Disturbance-adapted colonizer of bare ground; erosion control on slopes; tolerant of roadway salt and compacted soil.

Cultural Significance: Cypress spurge has been used medicinally as a laxative; its milky sap can cause dermatitis. It appears to have been introduced into North America an ornamental ground cover sometime in the mid-1800s.

Cypress
spurge in
a neglected
urban
landscape

Cypress spurge in bloom

Cypress spurge flowers

Cypress spurge produces milky
sap when injured

Cypress spurge foliage

Coronilla varia L.   Crownvetch

Synonyms: *Securigera varia,* axseed, trailing crownvetch

Life Form: herbaceous perennial; up to 3 feet (1 m) across

Place of Origin: Eurasia

Vegetative Characteristics: Crownvetch produces numerous sprawling stems from a long-lived, perennial base (the crown). The alternate, pinnately compound leaves are 2–6 inches (5–15 cm) long and consist of 9–25 pairs of small, oval, bristle-tipped leaflets, each about 0.5 inches (1 cm) long.

Flowers and Fruit: The pealike flowers are pink and white, about 0.5 inch (1 cm) long, and pleasantly fragrant. They are arranged in circular clusters (umbels) at the end of long, leafless stalks that emerge from the leaf axils and are pollinated mainly by bees. A large plant can produce hundreds of flower clusters, making for a conspicuous display in June and early July. The erect seed pods are 2–3 inches (5–8 cm) long and subdivided into segments, each of which contains a hard, brown seed.

Germination and Regeneration: Seeds germinate readily; established plants produce numerous trailing stems and rhizomes from the long-lived, perennial crown, the source of the common name.

Habitat Preferences: Crownvetch grows best in sunny, well-drained soil.

Ecological Functions: Tolerant of roadway salt and compacted soil; "fixes" atmospheric nitrogen in symbiosis with *Rhizobium* bacteria.

Cultural Significance: Crownvetch was introduced from China in the early 1900s by the U.S. Department of Agriculture. It was widely planted during the 1960s–1980s along highway embankments, especially the interstates that were being constructed at the time. The plant escaped cultivation and has become increasingly common in a variety of sunny, dry habitats.

Similar Species: Bird vetch (*Vicia cracca*) is a climbing plant with leaves that terminate in a tendril and flowers that are purple rather than pink.

Crownvetch is common along highways

Delicate compound leaves of crownvetch

Crownvetch flowers and foliage

Crownvetch flowers

Close-up of crownvetch flowers

Lotus corniculatus L.   Birdsfoot Trefoil

Synonyms: cat's clover, crow toes, sheep-foot, hop o'my thumb, devil's claw, poor-man's alfalfa

Life Form: herbaceous perennial; up to 6 inches (15 cm) tall with a spread of 2 feet (60 cm).

Place of Origin: Eurasia

Vegetative Characteristics: This mat-forming plant produces alternate, nearly sessile, compound leaves with 5 small, elliptical leaflets: 3 at the tip (the source of the name trefoil) and 2 smaller ones at the base of the stalk that look like stipules. Stems trail along the ground and root at the nodes, eventually forming circular clumps that can reach 2 feet (60 cm) in diameter and up to a foot (30 cm) tall.

Flowers and Fruit: Bright yellow pealike flowers are produced from June through summer and into early fall; 3–10 flowers are arranged in a flat cluster at the end of a 2–4 inch (2.5–5 cm) stalk that puts them well above the level of the foliage. They are pollinated by bees, and the pods that follow are about an inch (2.5 cm) long with a radial arrangement that makes them look vaguely like a bird's foot.

Germination and Regeneration: Seeds germinate readily in spring; established plants produce new shoots from a perennial crown; stolons and rhizomes develop in the fall.

Habitat Preferences: Because of its ability to "fix" atmospheric nitrogen, birdsfoot trefoil grows well in nutrient-poor, sandy soil and full sun. In the urban environment it is common in minimally maintained lawns, vacant lots, rubble dumps, and urban meadows. It is common along highway edges and median strips because it tolerates mowing and drought.

Ecological Functions: Tolerant of roadway salt and compacted soil; erosion control on slopes; soil building on degraded land.

Cultural Significance: This plant is widely planted for soil enrichment and stabilization in difficult landscape situations and as a nutritious forage crop for livestock. It was introduced into North America about 1900.

Similar Species: Crown vetch (*Coronilla varia*) is a sprawling plant that lacks tendrils and has flowers that are pink rather than white.

Birdsfoot trefoil growth habit in compacted soil

Birdsfoot trefoil flowering in early spring

Birdsfoot trefoil inflorescence

Nitrogen-fixing root nodules of birdsfoot

Birdsfoot trefoil seedpods (the source of the plant's common name)

Medicago lupulina L.   Black Medic

SYNONYMS: trefoil, black clover, none-such, hop medic

LIFE FORM: **winter** or **summer annual**; up to 6 inches (15 cm) tall with trailing stems

PLACE OF ORIGIN: Eurasia

VEGETATIVE CHARACTERISTICS: Black medic produces spreading branches from its base. The alternate leaves consist of 3 oval to wedge-shaped leaflets 0.4–0.8 inches (1–2 cm) long; they superficially resemble clover leaves, but the middle leaflet has a longer stalk than the lateral ones. The plant produces a shallow taproot and the stems do not produce adventitious roots.

FLOWERS AND FRUIT: Black medic produces dense, cylindrical heads of tiny, bright yellow, bee-pollinated flowers from May through September; the flower heads, which are about 0.5 inch (1 cm) wide, are carried on short stalks. The fruit is a curved, kidney-shaped black pod containing a single seed.

GERMINATION AND REGENERATION: Seeds germinate readily under a variety of disturbed conditions. Buried seeds retain their viability in the soil for many years.

HABITAT PREFERENCES: Black medic is highly drought tolerant and grows well on nutrient-poor soil in full sun. In the urban environment it is common in disturbed sites such as roadsides, compacted walkways, and waste areas.

ECOLOGICAL FUNCTIONS: Disturbance-adapted colonizer of bare soil; enriches the soil by "fixing" atmospheric nitrogen through a symbiotic relationship with root nodule–forming *Rhizobium* bacteria.

CULTURAL SIGNIFICANCE: Black medic was introduced into North America as a forage crop for livestock. It escaped from cultivation and has become widely naturalized.

SIMILAR SPECIES: **Alfalfa** or **lucerne (*Medicago sativa* L.)** is a much larger species from Europe that can grow up to 3 feet (1 m) tall and produces violet-blue flowers in flattened umbels. It is widely grown as forage for livestock and enriches the soil through its nitrogen-fixing root nodules. Dioscorides included alfalfa in his first-century herbal, *De Materia Medica,* to promote appetite and stop bleeding. It has escaped from cultivation and is often found growing in open urban meadows. Its sprouted seeds are common in salad bars across North America and Europe.

Black medic foliage
and flowers

Black medic inflorescence

Black medic's spreading
growth habit

Alfalfa growth habit

Alfalfa flowers

Melilotus officinalis (L.) Lam.   Sweet Clover

SYNONYMS: *Melilotus alba,* sweet melilot, bokhara, tree clover

LIFE FORM: **biennial**; up to 8 feet (1.6 m) tall

PLACE OF ORIGIN: Eurasia

VEGETATIVE CHARACTERISTICS: Sweet clover is a tall, widely branched plant with smooth stems. The alternate compound leaves consist of 3 narrow leaflets, 0.5–1 inch (1.2–2.5 cm) long, with finely toothed margins; the center leaflet has a longer stalk than the lateral leaflets. One-year-old plants produce a thick taproot that stores nutrients for the plant's second-year growth spurt. The entire plant is usually dead by September, well before frost. When growing under optimal conditions it can reach shrublike stature.

FLOWERS AND FRUIT: Sweet clover produces upright clusters (racemes) of white or bright yellow, pealike flowers at the ends of all its branches; the flowers are about 0.5 inch (1 cm) long and are pollinated by honeybees and other insects. The small seedpods are black to gray and contain 1 or 2 seeds.

GERMINATION AND REGENERATION: The seeds germinate readily in sunny, disturbed sites; they can also survive burial in the soil for many years.

HABITAT PREFERENCES: Sweet clover grows best in full sun and dry, sandy or gravelly soil. It is common on roadsides, vacant lots, rubble dumps, urban meadows, unmowed highway banks and median strips, and railroad rights-of-way.

ECOLOGICAL FUNCTIONS: Soil improvement through nitrogen fixing; taproot breaks up compacted soil.

CULTURAL SIGNIFICANCE: Sweet clover was intentionally introduced from Europe as fodder for livestock, but too much of it can produce toxic side effects. It has a long history of medicinal use in Europe to treat ulcers, inflammation, and chest complaints. It has also been used to flavor cheese and tobacco, and it yields excellent honey.

RELATED SPECIES: **Yellow sweet clover,** which produces bright yellow flowers, was formerly considered a distinct species from **white sweet clover,** but both are now treated as *Melilotus officinalis.*

White sweet clover growth habit

White sweet clover growing along a Boston street

White sweet clover flowers

Nitrogen-fixing nodules on white sweet clover roots

Yellow sweet clover growth habit

Yellow sweet clover flowers

Trifolium pratense L.   Red Clover

Synonyms: purple clover, meadow clover, peavine clover, cowgrass

Life Form: herbaceous biennial or **perennial**; up to 18 inches (50 cm) tall

Place of Origin: Eurasia

Vegetative Characteristics: Red clover is a semierect plant that produces numerous hairy stems from its base; the leaves are finely hairy and consist of a long petiole topped with 3 elongated leaflets—up to 1.5 inches (4 cm) long—each with a distinctive white to green, V-shaped chevron on the upper surface. At night the leaflets typically droop down, probably to conserve heat, water, or both. Mature plants develop a tough, fibrous root system. Red clover is typically a short-lived, clump-forming plant as opposed to white clover, which is a long-lived ground cover.

Flowers and Fruit: The pink, red, or magenta flowers are arranged in dense, round clusters with 1 or 2 leaves at their base. There are anywhere from 50–200 individual flowers per head, and they are pollinated by a variety of insects (especially honeybees and bumblebees). The flowers are produced from May through October, and the typical legume seedpods develop within a month of fertilization.

Germination and Regeneration: Red clover reproduces readily from seed; established plants produce short rhizomes that give rise to new plants.

Habitat Preferences: Red clover grows in many different types of soil. Its ability to "fix" atmospheric nitrogen in symbiotic association with *Rhizobium* bacteria enhances its ability to flourish in poor soils. The plant is common in abandoned grasslands, neglected residential and commercial landscapes, vacant lots, rubble dump sites, and unmowed highway banks and median strips. It is moderately tolerant of mowing, but not as tolerant as white clover.

Ecological Functions: Disturbance-adapted colonizer of bare ground; important nectar source for various pollinating insects; food for wildlife; soil improvement.

Cultural Significance: Red clover has long been cultivated in North America as a cover crop to recondition worn-out soil and as fodder for livestock, especially when grown in combination with timothy grass. A tea made from red clover flowers was traditionally used as a cough suppressant, a mild sedative, and a blood purifier; applied externally it was used to treat ulcers, corns, and burns. Red clover is an important honey plant for bees and is the state flower of the state of Vermont.

Related Species: Alsike clover (*Trifolium hybridum* L.) is an erect or ascending perennial from Europe that grows to about 2 feet (60 cm) tall. It produces sweetly fragrant, pinkish white flowers in a round head at the end of a long stalk; its trifoliate leaves lack the white markings typical of other clover species. Its seeds are capable of germinating after long periods of burial in the soil.

Red clover
growth habit

Red clover inflorescences

Close-up of red clover flowers

A trio of clovers: alsike (*left*), white (*center*),
and red (*right*)

Alsike clover
flowers

Trifolium repens L.   White Clover

SYNONYMS: Dutch white clover, honeysuckle clover, white trefoil

LIFE FORM: herbaceous perennial; up to 6 inches (15 cm) tall

PLACE OF ORIGIN: Eurasia

VEGETATIVE CHARACTERISTICS: White clover is a mat-forming species with smooth (hairless) stems and leaves. The leaves are alternate and consist of a long petiole with 3 rounded leaflets up to an inch (2.5 cm) long; they are dark green on the upper surface, gray-green below, and are marked with a light green or white, V-shaped chevron at their base. The leaflets typically droop down at night, probably to conserve heat, water, or both.

FLOWERS AND FRUIT: White clover produces its white flowers throughout summer in round heads up to an inch (2.5 cm) wide on long stalks. They are pollinated by insects (especially honey and bumblebees) and persist on the plant in a dry state after the fruits have formed. The pods are about 0.25 inch (5 mm) long and contain 3–6 yellow to brown seeds.

GERMINATION AND REGENERATION: White clover reproduces readily from seed, and established plants produce stolons that spread rapidly across the surface of the ground, rooting as they go. White clover seeds are also capable of germinating after long periods of burial in the soil.

HABITAT PREFERENCES: White clover is a highly adaptable plant that grows well in full sun in a variety of soil types. Because of its tolerance of mowing, white clover is a common component of most minimally maintained lawns in residential, commercial, and public landscapes. In the urban environment it is common along roadway margins, stone walls, and rock outcrops; and in vacant lots, rubble dumps, and urban meadows.

ECOLOGICAL FUNCTIONS: Disturbance-adapted, ground-covering colonizer; improves soil quality by fixing nitrogen; important nectar source for various flower-pollinating insects; browse food for wildlife.

CULTURAL SIGNIFICANCE: White clover has been used as a nutritious forage crop in North America since the 1700s, usually growing in combination with grasses. By fixing atmospheric nitrogen through its symbiotic relationship with *Rhizobium* bacteria, clover can improve the condition of low-nutrient soils. Numerous cultivars have been selected including 'Ladino', which is much larger than the wild type. Before broadleaf herbicides were developed, white clover was accepted as a normal component of every lawn. The plant often produces leaves with 4 leaflets in early spring and early fall, when day length is changing rapidly.

RELATED SPECIES: Rabbitfoot Clover (*Trifolium arvense* L.) is a European annual that can grow up to a foot (30 cm) tall; all of its parts, including the trifoliate, linear leaves, are covered with soft hairs. The plant is most conspicuous in late summer, when it produces upright clusters of pale pink to light gray flowers surrounded by silky-fringed sepals, giving them a distinctly fuzzy appearance reminiscent of a rabbit's foot. The species is extremely drought tolerant and is common in sandy soils along the edges of roadsides and walkways.

A planting of mostly white clover

Nitrogen-fixing nodules on white clover roots

White clover growth habit

White clover flowers

Rabbitfoot clover growth habit

Close-up of rabbitfoot clover flowers

Vicia cracca L. Bird Vetch

SYNONYMS: cow vetch, blue vetch, catpea, tinegrass

LIFE FORM: herbaceous perennial vine, up to 3–6 feet (1–2 m) long

PLACE OF ORIGIN: Eurasia

VEGETATIVE CHARACTERISTICS: The thin, weak stems of bird vetch either climb over other vegetation or sprawl across the ground to form a dense mat. The leaves are alternate and pinnately compound, with 8–12 pairs of linear leaflets, each 0.5–1 inch (1–3 cm) long, and a terminal tendril that facilitates climbing.

FLOWERS AND FRUIT: The individual pealike flowers are deep bluish purple, 0.5 inch (1 cm) long, and clustered along one side of racemes that are up to 4 inches (10 cm) long. Insect-pollinated flowers are produced from June through August and open in sequence from the bottom upward. At maturity the seedpods split into 2 twisted halves, explosively ejecting their seeds into the surrounding environment.

GERMINATION AND REGENERATION: Bird vetch reproduces readily from seed; established plants produce new stems from a perennial crown.

HABITAT PREFERENCES: This adaptable plant does equally well in sun or shade in a variety of soil types. It is common in neglected ornamental landscapes, vacant lots, rubble dumps, urban meadows, and unmowed highway banks and median strips.

ECOLOGICAL FUNCTION: Disturbance-adapted colonizer of bare ground.

CULTURAL SIGNIFICANCE: Originally introduced as a fodder crop for livestock, bird vetch has escaped cultivation to become an important component of most urban meadows.

Bird vetch
growth habit

Bird vetch inflorescences

Bird vetch foliage with tendrils at end of its leaves

Bird vetch coming up through a yew hedge

Bird vetch flowers

Galeopsis tetrahit L.   Hemp Nettle

SYNONYMS: dog nettle, wild hemp, flowering nettle

LIFE FORM: **summer annual**; up to 2 feet (70 cm) tall

PLACE OF ORIGIN: Europe

VEGETATIVE CHARACTERISTICS: Hemp nettle stems are conspicuously square, densely covered with bristly hairs, and have distinct swellings just below the leaf nodes. The leaves are opposite, 1–5 inches (3–12 cm) long, egg shaped, and have coarse teeth along their margins and long stalks. The plant branches freely from the lower leaf axils and can develop a broad, spreading growth form.

FLOWERS AND FRUIT: Hemp nettle produces clusters of flowers in terminal leafy spikes or in the axils of the upper leaves; the individual flowers, which are about 0.5 inch (1 cm) wide, can either be white or pale pink with magenta markings that guide insect pollinators. The calyx that envelops the developing fruit is armed with 5 sharp, prickly spines that break off when the plant is handled. These spines and the densely hairy stems are the source of the common name.

GERMINATION AND REGENERATION: The seeds germinate under a wide variety of conditions but seem to prefer moist soil in either sun or shade. Buried seeds remain viable for many years.

HABITAT PREFERENCES: Hemp nettle grows best in moist soil in full sun but will tolerate dry, shady conditions. In the urban environment it is common at the base of stone walls and rock outcrops, on the edges of minimally maintained lawns, and in disturbed waste places and vacant lots. It often forms dense stands.

ECOLOGICAL FUNCTION: Disturbance-adapted colonizer of bare ground.

RELATED SPECIES: Motherwort (*Leonurus cardiaca* L.) is a perennial member of the mint family with square stems and pairs of opposite leaves that are perpendicular to one another. It grows best in moist, nutrient-rich soil, but can also be found on dry, compacted sites in partial shade. Under optimal conditions the plant can be 5–6 feet (1.6–2 m) tall with numerous side branches and a tenacious root system. The uppermost leaves that subtend the flower clusters are narrow, and most have 3-pointed lobes; the lower leaves, which resemble maple leaves, are much broader and have 3–5 sharp-pointed lobes. The pink-purple, insect-pollinated flowers are produced in tight clusters in the axils of the upper leaves and occupy the top 12 inches or so (30–40 cm) of the plant's height. Motherwort has been used to treat a variety of "female complaints" and to relieve stress at least since the time of Dioscorides, who included it in his first-century herbal, *De Materia Medica*.

Hemp nettle growth habit

Hemp nettle flowers

Hemp nettle
in flower

Hemp nettle with
mature seed

Motherwort growth habit

Motherwort flowers

Glechoma hederacea L.   Ground Ivy

SYNONYMS: *Nepeta hederaceae, Nepeta glechoma,* gill-over-the-ground, creeping Charlie, cats-foot, ale hoof, gillale

LIFE FORM: perennial; stems up to 2 feet (70 cm) long

PLACE OF ORIGIN: Eurasia

VEGETATIVE CHARACTERISTICS: Ground ivy is a prostrate plant with square stems that root at the nodes. The leaves are shiny, kidney shaped, opposite, about an inch (2.5 cm) wide, and have slightly depressed veins and broadly rounded teeth along their margins. The foliage often has a purplish cast in early spring. In mild winters the leaves are evergreen. Under semi-shaded conditions, it can become a dominant groundcover.

FLOWERS AND FRUIT: Ground ivy produces clusters of purplish blue flowers in the upper leaf axils of specialized stems that are shorter and more upright than the trailing vegetative stems. The funnel-shaped, insect-pollinated flowers have 2 distinct lips, are about 0.5 inch (1 cm) long, and are produced from April through July. The tiny, dry fruits are enclosed in a hairy calyx.

GERMINATION AND REGENERATION: Ground ivy will germinate from seeds, but more typically it spreads by means of its creeping stems, which root at the nodes and can form large patches.

HABITAT PREFERENCES: Ground ivy grows best in shady, moist soil but can tolerate full sun and some drought. It is common in trampled or minimally maintained lawns; neglected residential, commercial, and municipal landscapes; on the margins of freshwater wetlands, ponds, and streams; and in highway drainage ditches.

ECOLOGICAL FUNCTIONS: Disturbance-adapted colonizer of bare ground; erosion control on slopes; food and habitat for wildlife.

CULTURAL SIGNIFICANCE: Ground ivy has a multiplicity of medicinal uses in Europe that date back at least to the first century when Dioscorides included the plant in first-century herbal, *De Materia Medica.* At one time, ground ivy was used instead of hops in fermenting and clarifying beer and ale. Traditionally, a tea made from the leaves has been used to treat diseases of the lungs and kidneys, asthma, and jaundice. The Shakers used it to treat chronic lead poisoning caused by the ingestion of paint. Ground ivy was an early arrival in North America, having been listed as a cultivated plant in Josselyn's *New-England's Rarities* (1672).

Ground ivy
in flower

Ground ivy produces long runners

Ground ivy foliage

Flowering ground ivy in a lawn

Close-up of ground ivy flowers

Lamium amplexicaule L.   Henbit

SYNONYMS: dead nettle, blind nettle, bee nettle, giraffe head

LIFE FORM: **winter annual**; up to 18 inches (40 cm) tall

PLACE OF ORIGIN: Europe

VEGETATIVE CHARACTERISTICS: The distinctive square stems of henbit can be either green or purple, and its growth habit can be either prostrate or erect. The dull green, opposite leaves are 0.5–0.75 inch (1–2 cm) long are more or less heart shaped with rounded teeth. The lower leaves produce a distinct petiole; the upper leaves encircle the stem with their base. Under sunny, moist conditions the plant can form large, showy patches.

FLOWERS AND FRUIT: In early spring henbit produces whorled clusters of showy pink to purple flowers in the axils of the uppermost leaves. The flower is a 2-lipped tube about 0.5 inch (1 cm) long that is pollinated by bees and other insects. The fruits that follow are enclosed within a small cup formed by the fused sepals.

GERMINATION AND REGENERATION: The seeds typically germinate in late summer or fall; buried seeds remain viable for many years.

HABITAT PREFERENCES: Henbit grows best during the cool weather of spring and fall in moist, fertile soil in full sun. In the urban environment it is common in minimally maintained lawns, ornamental planting beds, topsoil stockpiles, moist wetland edges, and roadside drainage ditches.

ECOLOGICAL FUNCTION: Disturbance-adapted colonizer of moist soil.

CULTURAL SIGNIFICANCE: Both henbit and purple dead nettle (see below) had limited use in traditional European medicine to stop bleeding and to promote perspiration. The young shoots are edible in spring.

RELATED SPECIES: **Purple deadnettle (*Lamium purpureum* L.)** is another European winter annual that produces light purple, insect-pollinated flowers in early spring. Unlike henbit, all of its leaves have distinct petioles; the upper leaves, which are crowded at the ends of the branches, are conspicuously reddish purple.

Henbit
flowers

Henbit in flower in early spring

Close-up of
henbit flowers

Purple
deadnettle
dies down in
mid- to late
spring

Close-up
of purple
deadnettle
flowers

Conspicuous
red leaves
of purple
deadnettle

235

Prunella vulgaris L.   Healall

Synonyms: *Brunella vulgaris*, self-heal, carpenter's weed, sicle-wort

Life Form: **evergreen perennial**; between 2 and 12 inches (5–30 cm) tall

Place of Origin: Europe, Asia, and North America

Vegetative Characteristics: Like most mints, healall has distinctly square stems. The opposite, lance-shaped leaves have distinct petioles and are 1–4 inches (2.5–10 cm) long depending on the growing conditions. The stems creep along the ground and root at the nodes.

Flowers and Fruit: From June through September healall produces dense, cylindrical spikes of flowers, up to 2 inches (5 cm) long, at the ends of upright stems. They are arranged in tiers or whorls and composed of a ring of 6 stalkless, light blue to purple, tubular flowers supported by a pair of spreading, sharp-pointed bracts. The flowers are pollinated by a wide variety of insects. The small fruits are embedded within the leafy calyx and contain a single brown seed.

Germination and Regeneration: Healall reproduces by seed and by creeping stems that root at the nodes.

Habitat Preferences: Healall grows in a wide variety of habitats from sun to shade and moist to dry. It tolerates close mowing and is abundant in minimally maintained lawns. It is also found in recently cleared woodlands, drainage ditches along highways, rock outcrops, and stone walls. In its native habitat healall grows in open grasslands and meadows.

Ecological Function: Disturbance-adapted colonizer of bare ground.

Cultural Significance: Healall is widely distributed throughout the temperate zones of the Northern Hemisphere. It has a long history of use as a medicinal plant wherever it is found, especially in Europe and China, where tea made from the fruiting heads has been used as a general tonic as well as to treat liver and circulation problems. In North America the Shakers sold healall to treat hemorrhages and sore throat and to increase the flow of urine.

Growth habit
and flowers of
unmowed healall

Healall inflorescence

Healall
growing
in a low-
maintenance
lawn

Close-up of healall flowers in a mowed
lawn

Healall flowers

Lythrum salicaria L.   Purple Loosestrife

SYNONYMS: purple lythrum, bouquet-violet, spiked loosestrife

LIFE FORM: **herbaceous perennial**; up to 5 feet (1.5 m) tall

PLACE OF ORIGIN: Eurasia

VEGETATIVE CHARACTERISTICS: The tall, square stems of purple loosestrife have lance-shaped to linear leaves that are 1–4 inches (3–10 cm) long and an inch (2.5 cm) wide; they are arranged opposite to one another or in whorls of 3.

FLOWERS AND FRUIT: Purple loosestrife produces terminal spikes up to 16 inches (40 cm) long with conspicuous pink-purple flowers that have 5–7 petals. Blooming, which extends from July through September, begins at the bottom of the spike and moves upward as the season progresses. Following pollination by nectar-feeding insects (mainly bees), numerous small capsules develop, each containing about 100 small, red-brown seeds. A large plant can produce thousands of seeds, which are dispersed by flowing water or on the feet of wading birds.

GERMINATION AND REGENERATION: Purple loosestrife reproduces freely from seeds, which can remain viable in the soil for many years. Established plants produce new shoots from a perennial crown, root suckers, or detached stems, all of which contribute to the formation of vast, dense stands.

HABITAT PREFERENCES: Purple loosestrife tolerates a wide range of soil moisture, pH, and nutrient conditions and can dominate the sunny margins of freshwater wetlands and meadows. It is common in brackish marshes and along the edges of ponds, streams, and rivers, especially those that have been contaminated with road salt and have a higher than normal pH. Loosestrife grows best on soil that stands just a few inches above the water table.

ECOLOGICAL FUNCTIONS: Nutrient absorption in wetlands; tolerant of roadway salt; stream and river bank stabilization.

CULTURAL SIGNIFICANCE: Purple loosestrife was introduced into North America in the early 1800s and is now widespread across much of the continent. It was probably introduced several times: unintentionally through ship's ballast or attached to sheep's wool, and intentionally for ornamental or medicinal purposes. A tea made from the entire flowering plant is a European folk remedy for diarrhea and dysentery, and a wash for cleaning wounds. The wetlands where this species is dominant are a spectacular sight in July and August when the plants are in bloom. Conservationists despise purple loosestrife, despite its beauty, and it is listed as an invasive species in most of the states where it grows. Numerous horticultural varieties are in cultivation. Charles Darwin made a detailed study of the plant's unusual floral morphology.

Purple loosestrife flowers

Purple loosestrife along the Concord River in Concord, Massachusetts

Purple loosestrife in full flower in August

Dead stems of purple loosestrife in winter

Purple loosestrife seedpods

Malva neglecta Wallr.   Common Mallow

Synonyms: *Malva rotundifolia*, cheeses, cheeseweed, cheese mallow, running mallow, round-leaved mallow, buttonweed

Life Form: **annual** or **biennial**; with stems up to 3 feet (1 m) long

Place of Origin: Eurasia

Vegetative Characteristics: Under urban conditions, common mallow is a low-growing, spreading plant with numerous stems that trail along the ground with their tips turned upward. Its alternate leaves are composed of a long, grooved petiole and a roughly circular blade, which can be 0.75–2.5 inches (2–6 cm) wide. The leaf has shallow lobes and prominent veins that radiate out from the point where the petiole is attached. Common mallow produces a deep-growing, tenacious taproot.

Flowers and Fruit: From late spring through early fall, common mallow produces small flowers that are about 0.5 inch (1.5 cm) across with 5 white to pale lavender petals that are notched at the tip. The flowers emerge at the point where the leaf is attached to the stem (the axil) and typically do not grow above the level of the foliage. They can either be insect- or self-pollinated. The common name "cheeses" derives from the small, flattened circular fruit with scalloped edges, which resembles a miniature wheel of cheese.

Germination and Regeneration: Seeds germinate readily on bare ground in sun or shade, and can survive burial in the soil for many years.

Habitat Preferences: Common mallow grows best in full sun and rich soil; it can also be found in compacted lawns, neglected ornamental landscapes, vacant lots, roadsides, and walkways.

Environmental Function: Disturbance-adapted colonizer of bare soil.

Cultural Significance: The young leaves and shoots are edible and because of their mucilaginous properties have been used to thicken soups. A tea made from the leaf or root has been used in traditional medicine to treat digestive problems, angina, and bronchitis. In the not too distant past, children ate the young green fruits, or "cheeses."

Common mallow growing in compacted soil

Common mallow in flower

Common mallow foliage

Close-up of common mallow flower

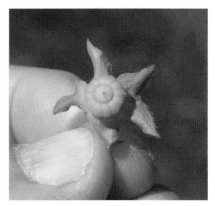

Common mallow fruits, or "cheeses"

Oenothera biennis L.   Evening Primrose

SYNONYMS: scabbish, tree primrose, fever plant, evening star, night willow-herb

LIFE FORM: herbaceous biennial; up to 6 feet (2 m) tall

PLACE OF ORIGIN: North America

VEGETATIVE CHARACTERISTICS: The first-year plant is a basal rosette of elliptic to lance-shaped leaves, 4–8 inches (10–20 cm) long, with distinct white mid-veins and coarsely toothed, wavy margins. The plant sends up a flowering stem during the second year of growth, typically with secondary branches arising from its midpoint. The leaves become progressively smaller towards the top of the stem.

FLOWERS AND FRUIT: Evening primrose produces conspicuous bright yellow flowers in the upper leaf axils from June through September. They are 0.75–2 inches (2–5 cm) in diameter with 4 large petals and fused sepals with tips reflexed downward. The flowers open toward the end of the day, emitting a fragrance that attracts night-flying moths. The woody seed capsules are about 1.5 inches (4 cm) long, slightly curved, and conspicuously persist on the dead stalks through winter.

GERMINATION AND REGENERATION: Seeds germinate readily in a wide variety of disturbed sites; seeds buried in the soil retain their viability for many years and will germinate when brought to the surface.

HABITAT PREFERENCES: Evening primrose grows best on dry, sandy or gravelly soil in full sun. It is common in neglected residential and commercial landscapes, minimally maintained public parks, vacant lots, rubble dumps, abandoned grasslands, meadows, small pavement openings and cracks, chain-link fence lines, unmowed highway banks and median strips, and railroad rights-of-way.

ECOLOGICAL FUNCTIONS: Disturbance-adapted colonizer of bare ground.

CULTURAL SIGNIFICANCE: This species appears to be native to most of eastern and central North America. It was introduced into Europe soon after North America was colonized and is now fully naturalized on that continent. The taproot is edible if the cooking water is changed twice. Evening primrose has been used in traditional herbal medicine to treat coughs, asthma, skin diseases, and "female complaints."

Evening primrose in bloom in Boston

Evening primrose in early summer

Evening primrose inflorescence

Evening primrose with spreading growth habit

Evening primrose rosette at the end of its first year of growth

Evening primrose seedpods

Oxalis stricta L.   Yellow Woodsorrel

Synonyms: *Oxalis europaea, Xanthoxalis stricta*, sour grass

Life Form: summer annual; up to 18 inches (50 cm) tall

Place of Origin: Europe and North America

Vegetative Characteristics: Yellow woodsorrel is a slender, erect or prostrate plant that produces weak, hairy, green or purple stems. The cloverlike leaves consist of 3 heart-shaped leaflets—0.4–0.8 inches (1–2 cm) wide—that fold up at night and in the heat of the noonday sun; they are arranged alternately and have long petioles.

Flowers and Fruit: The bright yellow flowers appear within a month of germination, and the plant continues producing them throughout the summer. The flowers are 0.25–0.5 inch (6–12 mm) wide, have 5 petals, and are produced in clusters on long stalks that emerge from the leaf axils. Although they are visited by insects, the flowers are mainly self-pollinated. The fruit is a 5-sided capsule that tapers to a sharp point and is about 0.5 inch (1–1.5 cm) long. It explodes when mature, ejecting the seeds up to 10 feet (3 m).

Germination and Regeneration: The seeds germinate readily in moist, nutrient-rich soil and are capable of germinating after long periods of burial. The plant also spreads by rhizomes that can be up to 6 inches (15 cm) long.

Habitat Preferences: Yellow woodsorrel grows best in sunny locations and moist, nutrient-rich soil but tolerates both shade and drought. It is common in neglected ornamental landscapes, minimally maintained lawns, vacant lots, rubble dumps, small pavement openings and cracks, rock outcrops, stone walls, drainage ditches, and median strips. In its native habitat it grows in grasslands and open woodlands.

Ecological Function: Disturbance-adapted colonizer of bare ground.

Cultural Significance: Yellow woodsorrel is widely distributed across the temperate world. Its leaves are edible but have a sour taste due to the oxalic acid they contain. Darlington's 1859 report that "the leaves of this very common plant have an agreeable acidity and are frequently eaten by children" is still true today.

Yellow woodsorrel
in flower and fruit

Long white
rhizomes
of yellow
woodsorrel

Yellow woodsorrel foliage

Yellow
woodsorrel
flower and
foliage

Yellow woodsorrel seedpods ready to explode

Chelidonium majus L.   Greater Celandine

SYNONYMS: swallowwort, tetterwort

LIFE FORM: **herbaceous biennial** or **short-lived perennial**; up to 3 feet (90 cm) tall

PLACE OF ORIGIN: Eurasia

VEGETATIVE CHARACTERISTICS: During its first year greater celandine produces a rosette of pinnately compound, irregularly lobed or coarsely toothed leaves that are 3–5 inches (8–13 cm) long and bright green on their upper surface and whitish on the underside. When broken the leaves and stems exude bright orange sap that dramatically stains the hands. First-year plants, which can be about 12 inches (30 cm) across and half as tall, are evergreen through the winter. During the second year plants elongate to produce hollow flower stalks with alternately arranged leaves; these stems are conspicuously swollen at the nodes and easily broken.

FLOWERS AND FRUIT: Throughout the spring and summer greater celandine produces numerous clusters of bright yellow flowers with 4 shiny petals at the ends of leafy stalks; they can either be insect- or self-pollinated. The slender, upright fruit capsules are about 1.5 inches (4 cm) long and explode when ripe, scattering their seeds in all directions. The entire plant dies in late summer or early fall.

GERMINATION AND REGENERATION: The seeds germinate readily in early spring in a variety of disturbed sites. Buried seeds remain viable for many years.

HABITAT PREFERENCES: Greater celandine grows best in sunny, moist, nutrient-rich soil but can tolerate shady, dry conditions. It is common in minimally maintained public parks; vacant lots and rubble dumps; the margins of freshwater wetlands, ponds, and streams; stone walls and rock outcrops; and neglected ornamental planting beds.

ECOLOGICAL FUNCTION: Disturbance-adapted colonizer of bare ground.

CULTURAL SIGNIFICANCE: Greater celandine has a long history of medicinal use. Dioscorides included it in his first-century herbal, *De Materia Medica*, and Josselyn reported in 1672 that celandine was cultivated in New England as a medicinal plant for treating various skin diseases. The orange sap, which can cause skin irritation, has been used in Europe and China to treat warts. In *American Weeds and Useful Pants* (1859) Darlington noted that greater celandine was "a common weed about dwellings. Its very brittle stems, when broken, exude a saffron-colored strong-smelling juice, which is very bitter and acrid. The plant was at one time much extolled as a remedy for jaundice, but little use is made of it, except that the fresh juice is occasionally applied to warts."

Greater
celandine
growth habit

Greater celandine growing on an old stone
bridge in Boston

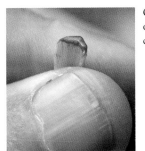

Orange sap
of greater
celandine

Developing fruits of greater celandine

Greater celandine flowers

Phytolacca americana L.   Pokeweed

SYNONYMS: *Phytolacca decandra*, poke salad, poke berry, poke root, skoke, pigeon berry, garget, ink berry, coakum, cancer jalap, red-ink plant, etc.

LIFE FORM: **herbaceous perennial**; up to 6 feet (2 m) tall

PLACE OF ORIGIN: eastern North America

VEGETATIVE CHARACTERISTICS: The simple, alternate leaves of pokeweed are broadly lance shaped, 4–12 inches (10–30 cm) long, and about a third as wide. They are pale green with a smooth, waxy surface, and typically wilt in the noonday sun. Under the right conditions poke can grow into a large bush that dies back to a thick perennial taproot with the first frost. The fleshy, hollow stems take on a dramatic reddish purple color in late summer and fall.

FLOWERS AND FRUIT: Pokeweed produces nodding to erect racemes of flowers from July through September. The racemes are 4–8 inches (10–20 cm) long, and the small individual flowers have 5 white or green "petals." The flowers are mainly self-pollinated and quickly mature into long chains of deep black-purple fruits that stain the hands and are highly attractive to both birds and children.

GERMINATION AND REGENERATION: Birds eagerly consume mature pokeweed fruits in the fall, and the seeds germinate readily the following spring. Seeds that have been buried in the soil for many years will germinate when brought to the surface by some form of disturbance. Established plants produce new stems from the perennial, taproot which can be up to a foot (30 cm) long by 4 inches (10 cm) wide.

HABITAT PREFERENCES: Pokeweed grows best in deep, moist soils in full sun and can live for several years under such conditions. It readily colonizes freshly disturbed ground by means of buried seeds and is common in neglected ornamental landscapes; vacant lots and rubble dumps; freshwater wetlands, ponds, and streams; chain-link fence lines; unmowed highway banks; the understory of disturbed woodlands; and railroad rights-of-way.

ECOLOGICAL FUNCTIONS: Food and habitat for wildlife; erosion control on slopes; stream and river bank stabilization.

CULTURAL SIGNIFICANCE: In *American Weeds and Useful Plants* (1859), Darlington summed up pokeweed's virtues: "The young shoots of this plant afford a good substitute for Asparagus; the root is said to be actively emetic; and the tincture of the ripe berries is, or was, a popular remedy for chronic rheumatism. The mature berries have been used by the pastry cook in making pies of equivocal merit. Notwithstanding all this, the plant is regarded and treated as a weed by all neat farmers." Red dye and ink can be made from the juice of the berries. Indeed, the first copy of the Declaration of Independence, which is now badly faded, was written with pokeweed ink. Pokeweed is considered an invasive species in many parts of Europe.

Pokeweed in the fall

Pokeweed has moved into this abandoned house

Pokeweed growth habit

Pokeweed inflorescence

Young pokeweed plant

Maturing pokeweed fruits

Plantago lanceolata L.   Buckhorn Plantain

SYNONYMS: English plantain, buckhorn, ribgrass, ribwort, black-jacks, narrow-leaved plantain, jackstraw, black jack

LIFE FORM: perennial; up to 18 inches (50 cm) tall

PLACE OF ORIGIN: Europe

VEGETATIVE CHARACTERISTICS: The smooth, lance-shaped leaves have prominent parallel veins (the so-called ribs) and are 2–10 inches (5–25 cm) long and less than an inch (2.5 cm) wide. The leaves are arranged in a basal rosette with a slight twist and an erect orientation. New leaves and flowers grow up from a short underground stem that also produces a fibrous root system. In mild winters the leaves remain evergreen.

FLOWERS AND FRUIT: Buckhorn plantain blooms from June through September, producing short, dense spikes of greenish white, wind-pollinated flowers at the end of rigid, unbranched stalks. Individual flowers open from the bottom up in a spiral pattern, often producing a distinct ring of pollen-producing stamens at any point in time. The fruit is a small capsule containing 2 seeds.

GERMINATION AND REGENERATION: The seeds become sticky when wet, which facilitates their dispersal by animals. They germinate readily in spring in sunny locations. The plant can also sprout from root pieces left in the ground following unsuccessful attempts at removal.

HABITAT PREFERENCES: Buckhorn plantain grows best in sun and tolerates compacted, pH-neutral soils and close mowing. In the urban environment it is common in minimally maintained lawns, urban meadows, compacted pathways, vacant lots, rubble dumps, rock outcrops and stone walls, roadsides, median strips, and small pavement openings.

ECOLOGICAL FUNCTION: Disturbance-adapted colonizer of bare ground.

CULTURAL SIGNIFICANCE: Buckhorn plantain has a long history of medicinal use in European folk medicine. The leaves make a leaf tea used to treat coughs, diarrhea, and dysentery; the fresh leaves can be applied directly to blisters, sores, and ulcers to reduce the pain of inflammation. In recent years it has become an important research subject for studying the complex interaction between plants and their mycorrhizal fungi.

Buckhorn plantain in an unmowed field

Buckhorn plantain growing in a crack in a wall

Buckhorn plantain dominates this neglected lawn

Buckhorn plantain foliage

Buckhorn plantain inflorescences

Plantago major L.   Broadleaf Plantain

SYNONYMS: *Plantago asiatica*, dooryard plantain, white man's foot, way-bread, English plantain, ribwort, ripple grass

LIFE FORM: **semi-evergreen perennial**; up to 12 inches (30 cm) tall

PLACE OF ORIGIN: Europe

VEGETATIVE CHARACTERISTICS: The leaves are arranged in a basal rosette and are 2–7 inches (5–18 cm) long and up to 4 inches (10 cm) wide. They are elliptic to egg-shaped with smooth margins, parallel veins, and a broad petiole with edges that curve upward. Plucking the leaves exposes the long, stringy fibers that traverse the underside of the leaf and can be removed in their entirety.

FLOWERS AND FRUIT: From June through September broadleaf plantain produces numerous cylindrical flower stalks 4–12 inches (10–30 cm) tall that are crowded with hundreds of inconspicuous greenish white, wind-pollinated flowers. The fruit is an oval capsule that, at maturity, splits around its midpoint when releasing its numerous seeds.

GERMINATION AND REGENERATION: The seeds germinate readily in a wide variety of habitats.

HABITAT PREFERENCES: While broadleaf plantain grows best in moist, nutrient-rich soil, it is remarkably tolerant of compacted, dry soils and can even tolerate close mowing. In urban areas it is common in trampled lawns, neglected ornamental landscapes, vacant lots, urban meadows, drainage ditches, and compacted pathways. In its native habitat it grows on sunny cliffs and rocky areas.

ECOLOGICAL FUNCTION: Disturbance-adapted colonizer of bare ground.

CULTURAL SIGNIFICANCE: Native Americans reportedly called broadleaf plantain the "white man's footprint" because it grew where the Europeans cleared the land. Josselyn (1672) listed the plant under the category: "Of such plants as have sprung up since the English planted and kept cattle in New England." Broadleaf plantain has long been used in Europe to treat inflammations, fevers, and sores, and—applied externally—to stop bleeding (it is a powerful coagulant).

SIMILAR SPECIES: **Blackseed plantain (*Plantago rugelii* Decne.)** looks very similar to broadleaf plantain but is a native of North America rather than an introduced species. Its seeds are black rather than brown, and its fruit capsules open by splitting below their midpoint. Its leaves are generally larger, smoother and lighter green as well, and tend to be wavy margined with petioles that are reddish at the base. Both species are common in disturbed habitats.

Broadleaf plantain in a typical urban niche

Stand of broadleaf plantain in full bloom

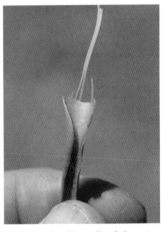

Leaf vessels of broadleaf plantain

Broadleaf plantain can tolerate being driven over

Blackseed plantain can dominate moist sunny sites

Polygonum aviculare L.   Prostrate Knotweed

SYNONYMS: knot-grass, door-weed, mat-grass, pink-weed, bird-grass, stone-grass, way-grass, goose-grass, allseed, centinode, nine-joints, wine's-grass

LIFE FORM: summer annual; up to 12 inches (30 cm) tall by 24 inches (60 cm) wide

PLACE OF ORIGIN: Introduced from Europe, possibly native to North America

VEGETATIVE CHARACTERISTICS: The alternate, gray-green leaves are lance shaped with pointed tips and tapered bases, and are between 0.5 and 1.25 inches (1–3 cm) long. Under stressful urban conditions the plant has small leaves and forms a wiry, prostrate mat that creeps along the ground (subspecies *depressum*); in good soil it develops into a more upright plant with much larger leaves (subspecies *aviculare*). As with all *Polygonum* species, a papery sheath (ocrea) encircles the stem at the leaf nodes or joints, giving them a "knotted" appearance. While this plant grows as a ground cover, it does not root at the nodes, instead focusing its energy on the development of a slender, tenacious taproot. In fall its leaves are often dusted with white splotches of powdery mildew.

FLOWERS AND FRUIT: Prostrate knotweed produces inconspicuous tiny white flowers with pink edges from June though September. They are typically self-pollinated and give rise to large numbers of equally inconspicuous fruits.

GERMINATION AND REGENERATION: The seeds germinate readily in early spring.

HABITAT PREFERENCES: Prostrate knotweed has a cosmopolitan distribution across the temperate world. It is ubiquitous in the urban environment and is most at home on hard, compacted soils that experience heavy pedestrian or vehicular traffic. It is common in the cracks that develop between granite curbs and the adjacent pavement, trampled lawns (it tolerates mowing), neglected ornamental landscapes, small pavement openings, rock outcrops, and stone walls. In its native European habitat prostrate knotweed grows on limestone cliffs, the natural analog to the pavement cracks it inhabits in the city.

ECOLOGICAL FUNCTIONS: Disturbance-adapted colonizer of bare ground; food for wildlife (especially birds).

CULTURAL SIGNIFICANCE: Prostrate knotweed was used in traditional European medicine to stop nosebleeds and to treat inflammations, wounds, and hemorrhoids. The plant was an early arrival in North America, having been listed by Josselyn in *New-England's Rarities* (1672) under the category: "Of such plants as have sprung up since the English planted and kept cattle in New England." By 1847 William Darlington reported that "this humble weed is thoroughly naturalized, and is one of the commonest everywhere about dwellings."

SIMILAR SPECIES: Spotted spurge, purslane, and carpetweed are other examples of mat-forming, taprooted plants with a circular growth pattern that seem "pre-adapted" to grow in sidewalk cracks and survive heavy foot traffic.

Prostrate knotweed growth form

Prostrate knotweed can grow just about anywhere

Prostrate knotweed is tolerant of compacted soil

Prostrate knotweed in its typical urban niche

Close-up of prostrate knotweed flowers

Polygonum cuspidatum Sieb. & Zucc.   Japanese Knotweed

SYNONYMS: *Fallopia japonica, Reynoutria japonica, Pleuropterus cuspidatus,* Japanese bamboo, Mexican bamboo

LIFE FORM: herbaceous perennial; up to 10 feet (3 m) tall

PLACE OF ORIGIN: temperate East Asia

VEGETATIVE CHARACTERISTICS: Japanese knotweed is a fast-growing, robust perennial that can form large clumps. It is often referred to as Japanese bamboo because its thick stems are hollow and jointed. The large, alternate leaves are broadly egg shaped, up to 6 inches (15 cm) long by 3–4 inches (7.5–10 cm) wide, and have a pointed tip and a prominent papery sheath (ocrea) where they attach to the stem.

FLOWERS AND FRUIT: In late summer Japanese knotweed produces highly conspicuous clusters of white flowers, 4–5 inches (10–13 cm) long, in the axils of its leaves. The species is dioecious with separate male and female individuals, and pollen is transferred by insects or wind. Female plants produce small, triangular, winged seeds that are dispersed by wind or water in late fall.

GERMINATION AND REGENERATION: The seeds germinate in a variety of conditions; established plants spread vigorously by producing new stems from long, deep-growing rhizomes. The plant can be spread inadvertently when topsoil containing the rhizomes is relocated.

HABITAT PREFERENCES: Japanese knotweed grows best on moist soil in full sun, but tolerates shade and drought. It is commonly found along the margins of freshwater wetlands, ponds, streams, and rivers (particularly where the banks are unstable); in vacant lots and rubble dumps; on unmowed highway banks; in roadside drainage ditches; and along fence lines and railroad tracks.

ECOLOGICAL FUNCTIONS: Tolerant of road salt and compacted soil; erosion control on slopes; stream and river bank stabilization; soil building on degraded land.

CULTURAL SIGNIFICANCE: Japanese knotweed was originally introduced into the United States as an ornamental plant during the 1870s and was actively sold by nurseries through at least 1910. The plant has escaped cultivation, and many states now list it as an invasive species. Once established it can be notoriously difficult to eradicate. The young shoots, which look a bit like asparagus, are edible after boiling and have a lemony taste. The roots contain large quantities of resveratrol, the same antioxidant compound found in red grapes and red wine that is thought to promote longevity. Indeed, Japanese knotweed is well on its way to becoming the most important commercial source of this popular "nutraceutical."

RELATED SPECIES: Japanese knotweed hybridizes readily with *Polygonum sachalinense,* a similar species from northeast Asia with leaves up to 14 inches (35 cm) long. The hybrid, *P. × bohemicum,* is intermediate in leaf characteristics and produces abundant viable seed.

Hollow stems of
Japanese knotweed

Japanese knotweed in flower in August

Japanese knotweed in early
spring

Japanese knotweed can form dense stands

Japanese knotweed in winter

Japanese knotweed fruits

Polygonum lapathifolium L.   Pale Smartweed

Synonyms: *Persicaria lapathifolia*, dockleaf smartweed, nodding smartweed

Life Form: summer annual; up to 4.5 feet (1.5 m) tall

Place of Origin: eastern North America and Europe

Vegetative Characteristics: Pale smartweed produces multiple smooth or slightly hairy stems, either green or red, from its base. The stems are noticeably swollen and jointed at the nodes, and the papery sheath (ocrea) that surrounds the stem at the base of the leaf lacks the fringe of fine hairs seen in *P. persica* and *P. caespitosum*. The purple-tinged leaves are alternate, lance shaped to elliptic, 2–6 inches (5–15 cm) long, and are often marked with a dark splotch.

Flowers and Fruit: White or pale pink to rose-colored flowers are produced from July through October. They are arranged in drooping or arching terminal spikes up to 3 inches (8 cm) long that lack sticky hairs; individual flowers are small and lack petals and can either be self- or insect-pollinated. The fruits are small and black.

Germination and Regeneration: The seeds germinate readily in sunny, disturbed sites.

Habitat Preferences: Pale smartweed grows best in moist, nutrient-rich soil in full sun but also grows in dry or compacted soil. In the urban environment it is common at the base of stone walls, in poorly drained areas of lawns and meadows, in vacant lots and waste areas, and on disturbed or compacted soil adjacent to wetlands.

Environmental Functions: Disturbance-adapted colonizer of bare ground; food and habitat for wildlife.

Cultural Significance: Native Americans used a tea made from the whole plant to treat diarrhea and a poultice of the leaves for hemorrhoids.

Related Species: Pennsylvania smartweed (*Polygonum pensylvanicum* L.), another native summer annual, resembles pale smartweed in its growth habit but produces its pale pink to rose-colored flowers in stubby, upright terminal spikes that are only about 1.5 inches (3 cm) long and are covered with sticky hairs. The lance-shaped leaves are sometimes marked in the middle with a darkly pigmented spot. In sunny, moist conditions and good soil, such as drainage ditches, the plant can be up to 4 feet (1.3 m) tall.

Pale smartweed growth habit

Pale smartweed produces a range of flower colors

Pale smartweed flowers are typically white

Pennsylvania smartweed growth habit

Unfringed ocrea of Pennsylvania smartweed

Pennsylvania smartweed flowers

Polygonum persicaria L.   Ladysthumb

Synonyms: *Persicaria persicaria*, heart's ease, smartweed, spotted knotweed, red-shank

Life Form: **summer annual**; up to 3 feet (90 cm) tall

Place of Origin: Europe

Vegetative Characteristics: Ladysthumb is a densely branched plant with smooth green or red stems that are swollen at the nodes to form distinct joints, or "knots." The leaves are lance shaped to elliptical; 1–6 inches (2.5–15 cm) long; and have a conspicuous darkly pigmented, triangular mark in the middle of the leaf blade (the so-called lady's thumbprint). The sheath surrounding the stem "joints" (ocrea) is covered with short bristles. Under favorable conditions ladysthumb will form large, spreading clumps.

Flowers and Fruit: Ladysthumb produces bright pink to red, and sometimes white, petal-less flowers from July through October. They are about an inch (2.5 cm) long and are arranged in dense, spikelike clusters at the ends of the branches; they can either be self- or insect-pollinated. The seeds are smooth and black.

Germination and Regeneration: The seeds germinate readily under a wide variety of conditions and can remain viable in the soil seed bank for many years.

Habitat Preferences: Ladysthumb grows best in moist soil in either sun or shade. Its size and vigor vary according to the growing conditions. It is common in neglected residential and commercial landscapes; minimally maintained public parks; vacant lots and rubble dumps; the margins of freshwater wetlands, ponds, and streams; small pavement openings and cracks; and unmowed highway banks, drainage ditches, and median strips.

Ecological Functions: Disturbance-adapted colonizer of bare ground; food for pollinating insects.

Cultural Significances: Ladysthumb has been used as a medicinal plant since at least the first century; Dioscorides included it in his first-century herbal, *De Materia Medica*. A tea made from the leaves was used in traditional European medicine as an astringent and to treat internal bleeding and menstrual problems. The plant was an early arrival in North America, having been listed in Josselyn's *New-England's Rarities* (1672).

Related Species: **Oriental ladysthumb (*Polygonum caespitosum* Blume)**, a native of East Asia that grows to about 1 foot (33 cm) tall, is very similar to *P. persicaria* but has much longer bristles on the ochrea, and the pigmented spot on its leaf is not quite as dark. Oriental ladysthumb flowers from June through the end of October. It often produces numerous stems from the base, leading to the formation of large, spreading clumps. It is common along the moist, sunny edges of sidewalks and roadways as well as in the heavily shaded understory of disturbed woodlands. *P. caespitosum* was inadvertently introduced into North America in the 1920s and has been rapidly increasing in abundance since then. In some parts of the urban environment it seems to be more common than *P. persicaria*.

Ladysthumb growth habit

Ladysthumb foliage

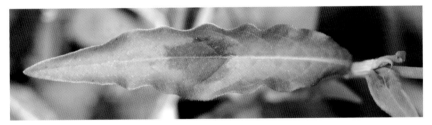

Ladysthumb leaf with "thumbprint"

Dense stand of Oriental ladysthumb

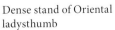

Long-bristled ocrea of Oriental ladysthumb

Ladysthumb flowers

Polygonum scandens L.   Climbing False Buckwheat

Synonyms: *Fallopia scandens,* climbing buckwheat, hedge smartweed, bindweed, winged bindweed

Life Form: herbaceous perennial vine; climbing to 15 feet (5 m)

Place of Origin: eastern North America

Vegetative Characteristics: Climbing false buckwheat climbs by means of thin, wiry, red stems that twine around other plants; in the absence of any support the plant will trail along the ground. The leaves are 1–3 inches (2–7 cm) long, alternate, triangular to heart shaped, and have 2 prominent basal lobes. A papery sheath (ocrea) surrounds the swollen stem at the base of each leaf.

Flowers and Fruit: Climbing false buckwheat produces white to greenish white or yellowish green flowers from July through October. The inflorescences (racemes) are 2–4 inches (5–10 cm) long and emerge from the upper leaf axils, standing above the foliage. The individual flowers are inconspicuous and, following pollination by insects, develop rapidly into upright stalks of green fruits with 3 distinct wings.

Germination and Regeneration: The wind-dispersed seeds germinate in the moist shade of other plants; established plants resprout from a perennial crown.

Habitat Preferences: Climbing false buckwheat grows well in a wide range of soil types and light conditions from full sun to full shade. In the urban environment it is common on chain-link fences and climbing up through planted shrubs.

Environmental Function: Disturbance-adapted colonizer of disturbed ground.

Related Species: Wild buckwheat or **black bindweed (*Polygonum convolvulus* L.)** is an annual vine from Europe with a growth habit similar to that of climbing false buckwheat, but its stems are only about 3 feet (1 m) long and its inflorescences are shorter—1–2 inches (2–5 cm) long—and have fewer flowers. Most notably for identification purposes, wild buckwheat fruits lack conspicuous wings.

Climbing false buckwheat in the urban environment

Conspicuous developing seeds of climbing false buckwheat

Climbing false buckwheat climbs by means of slender red stems

Climbing false buckwheat foliage

Climbing false buckwheat seeds have prominent wings

Rumex acetosella L.   Red Sorrel

SYNONYMS: sheep sorrel, sour grass, Indian cane, field sorrel, horse sorrel, sour weed, red weed, cow sorrel, sour dock

LIFE FORM: **semi-evergreen perennial**; up to 18 inches (45 cm) tall

PLACE OF ORIGIN: Eurasia and North Africa

VEGETATIVE CHARACTERISTICS: Red sorrel is a low-growing plant that produces a rosette of narrow, smooth leaves roughly 3 inches (7 cm) long and 1 inch (2.5 cm) wide with 2 prominent basal lobes that make them look like small arrowheads. Later in the growing season the plant produces upright flowering stalks with alternate, linear leaves that lack basal lobes. Under stressful growing conditions and at the end of the growing season the foliage takes on a deep reddish coloration. The slender, yellow roots are fibrous and freely branched.

FLOWERS AND FRUIT: Loose spikes, up to 18 inches (45 cm) tall, of inconspicuous, wind-pollinated flowers are produced from May through September. Yellow-green male flowers and reddish brown female flowers are produced on separate plants (dioecious). The reddish brown seeds mature in the fall and lack the prominent wings that characterize other *Rumex* species.

GERMINATION AND REGENERATION: Buried seeds can remain viable for many years and germinate when brought to the surface by some form of disturbance. Established plants spread vigorously by means of rhizomes and can form large colonies. The yellow roots can produce shoots when left behind after weeding. Large patches of red sorrel usually consist of only one sex, an indicator of asexual reproduction.

HABITAT PREFERENCES: Red sorrel grows best in full sun and is extremely tolerant of both acidic and poorly drained soils as well as drought. In the urban environment it is common in neglected ornamental landscapes, compacted lawns, urban meadows, small pavement openings, rock outcrops and stone walls, and roadsides. In its native habitat it grows in grasslands, disturbed open areas, and scree slopes.

ECOLOGICAL FUNCTIONS: Disturbance-adapted colonizer of acidic, low-fertility soils; food for wildlife and insects.

CULTURAL SIGNIFICANCE: High concentrations of oxalic acid give red sorrel leaves a sour taste and can sicken livestock that eat too much of the plant. European herbalists have used the plant as a diuretic to treat urinary and kidney problems, and in North America the Shakers sold it to treat skin diseases and as a poultice for boils and tumors. Red sorrel is closely related to cultivated sorrel (*Rumex acetosa*), and cooked leaves can be used as a base for purees.

Male clone of red sorrel in foreground with pink flowers; female clone with greenish flowers behind

Red sorrel foliage

Close-up of red sorrel foliage

Female red sorrel in bloom

Male red sorrel in bloom

Rumex crispus L.   Curly Dock

SYNONYMS: *Rumex elongatus,* sour dock, bitter dock, yellow dock, narrow dock, dock root, garden patience

LIFE FORM: herbaceous perennial; up to 5 feet (1.5 m) tall

PLACE OF ORIGIN: Europe

VEGETATIVE CHARACTERISTICS: Curly dock starts the growing season as a rosette of shiny narrow leaves up to a foot long (30 cm) and 2 inches (5 cm) wide with wavy, crumpled margins. In June the flowering stems elongate with alternate leaves; a papery sheath (ocrea) covers the point where the leaf attaches to the stem. The plant produces a stout, tenacious taproot.

FLOWERS AND FRUIT: Curly dock produces a tall flowering stalk in June and July that is covered with small, greenish, wind-pollinated flowers. The fruits, which turn deep brown at maturity, consist of a single seed attached to a papery, 3-winged structure (the calyx) that facilitates dispersal by wind or water. The distinctive chestnut brown seed stalks persist through the winter.

GERMINATION AND REGENERATION: Curly dock seeds germinate readily on bare ground; when buried they can remain viable for many years. Established plants resprout from a perennial crown; root fragments can give rise to new plants.

HABITAT PREFERENCES: Curly dock grows best on moist, heavy soils in full sun. In the urban environment it is common in neglected ornamental landscapes, vacant lots, rubble dumps, urban meadows, unmowed highway banks and drainage ditches, and railroad rights-of-way.

ECOLOGICAL FUNCTIONS: Disturbance-adapted colonizer of bare ground; food for wildlife.

CULTURAL SIGNIFICANCE: The root has a long history of use in European folk medicine as a laxative; an astringent; and a treatment for lung, liver, and skin problems. Curly dock was an early arrival in North America, having been listed in Josselyn's *New-England's Rarities* (1672) under the category: "Of such plants as have sprung up since the English planted and kept cattle in New England." The greens are edible in spring, after one or two changes of water.

RELATED SPECIES: Broadleaf dock (*Rumex obtusifolius* L.) is a European perennial that grows to be about 4 feet (1.3 m) tall. Its leaves are flatter and much broader than those of curly dock; they are heart-shaped at the base and lack a wavy margin. Broadleaf dock grows in many of the same nutrient-rich, sunny locations as curly dock, and the two plants are often seen growing side by side. Broadleaf dock was once a popular treatment for nettle stings.

Curly dock in
Tuscany, Italy

Curly dock growing in a drainage swale

Curly dock by the side of a Detroit
street

Broadleaf dock growth habit

Broadleaf dock foliage

Portulaca oleracea L.   Common Purslane

SYNONYMS: pussley, pursley, pressley, wild portulaca, little hogweed

LIFE FORM: summer annual; up to 6 inches (15 cm) tall by 2 feet (60 cm) wide

PLACE OF ORIGIN: a cosmopolitan plant of Eurasian origin

VEGETATIVE CHARACTERISTICS: Common purslane is a prostrate, mat-forming plant with smooth red or green stems that grow out to form a nearly perfect circle. The succulent, alternate leaves are about an inch (2.5 cm) long and are shaped like a wedge or a paddle—broad at the tip and narrow at the base. The leaves have a smooth, glossy surface that makes them glisten in bright sunlight. The plant forms a distinct taproot and its stems typically do not root at the nodes.

FLOWERS AND FRUIT: Purslane produces bright yellow flowers from July through September, roughly a month after germination; they have 5 short petals, lack a distinct stalk or petiole, and are clustered at the ends of the branches or in the leaf axils. Typically the flowers open only when it is sunny; they are mostly self-pollinated. The small, globe-shaped capsules are about 0.25 inch (8 mm) long and contain numerous tiny black seeds.

GERMINATION AND REGENERATION: Purslane is a "hot-weather weed" whose seeds typically do not begin germinating until the soil warms up in June. Seeds will continue sprouting throughout the summer, especially after rain. Plants that have been uprooted and discarded on the soil surface can use the water stored in their succulent stems and leaves to develop a new root system and continue growing. Buried seeds can remain viable for up to 40 years.

HABITAT PREFERENCES: Purslane grows best in nutrient-rich sandy soils in full sun but tolerates dry, compacted soil. It is common in neglected residential and commercial landscapes, vacant lots, rubble dumps, compacted or worn-out lawns, horticultural planting beds, and small pavement openings and cracks.

ECOLOGICAL FUNCTION: Disturbance-adapted colonizer of bare ground.

CULTURAL SIGNIFICANCE: Purslane is widely distributed throughout the temperate and subtropical regions of the world. The species name, *oleracea*, means "of the vegetable garden" and refers to the plant's edibility. Wherever purslane grows—either wild or cultivated—people have eaten its mucilaginous leaves and stems raw or have used them to thicken soups and stews. The leaves have a sharp, acidic taste and are a rich source of omega-3 fatty acids. The plant has a long history of medicinal use in Europe and Asia, mainly to reduce inflammation. Dioscorides included it in his first-century herbal, *De Materia Medica*. Purslane was being cultivated early as 1631 at John Winthrop's Plymouth colony in New England. Horticultural selections with leaves that are larger than normal or are reddish or yellowish are available in the nursery trade.

SIMILAR SPECIES: Spotted spurge, carpetweed, and prostrate knotweed are other examples of mat-forming plants with a circular growth pattern and a distinct taproot that seem "preadapted" to grow in sidewalk cracks and survive being walked on.

Purslane growth habit

Purslane dominating a patch of meadow

Purslane in its typical urban niche

Purslane foliage and flowers

Growth habit of a purslane seedling

Ranunculus bulbosus L.   Bulbous Buttercup

Synonyms: bulbous crowfoot, yellowweed, blister flower, gowan

Life Form: herbaceous perennial; up to 18 inches (45 cm) tall

Place of Origin: Eurasia

Vegetative Characteristics: Bulbous buttercup is a relatively low-growing plant with a basal rosette of 3-parted compound leaves; the middle leaflet has a distinct stalk and the 2 lateral leaflets are stalkless (sessile); the basal leaves have long, hairy petioles. The plant produces a distinct bulb about 0.5 inch (1.5 cm) wide at its base.

Flowers and Fruit: Shiny yellow flowers are produced at the ends of sparsely leaved, hairy stems from April through June. The insect-pollinated flower is about an inch (2.5 cm) wide with 5–7 petals that are rounded at the apex and wedge shaped at the base. The green sepals below the petals curve back against the stem (recurved). Each flower gives rise to a globe-shaped cluster of flattened, dry fruits.

Germination and Regeneration: The seeds germinate readily in spring or fall in sunny sites; in spring, established plants sprout from a marble-sized bulb.

Habitat Preferences: Bulbous buttercup is common in dry lawns; neglected residential and commercial landscapes; minimally maintained public parks and open spaces; vacant lots and rubble dump sites; freshwater wetlands, ponds, and streams; rock outcrops and stone walls; and unmowed highway banks and median strips with frequent salt applications.

Environmental Functions: Disturbance-adapted colonizing species; food for wildlife.

Cultural Significance: The leaves and stems of this plant and *R. acris* (see below) contain acrid juices that discourage cattle from browsing them, and the plants often become dominant in overgrazed pastures. Bulbous buttercup has had limited use as a treatment for arthritis and rheumatism because of its caustic properties. The bulbs are supposedly edible when collected in early spring and thoroughly dried. In the familiar childhood game, the reflected glow of the flower's yellow petals held under the chin identifies a lover of butter.

Related Species: Tall buttercup (*Ranunculus acris* L.) is taller than bulbous buttercup—reaching up to 3 feet (1 m)—and lacks the distinct bulb at the base of the stem. The large, hairy leaves are deeply divided into 3–7 lobes; the central lobe lacks a distinct stalk. The flowers of the two species are similar, but the sepals of tall buttercup are relatively small and inconspicuous, and are not recurved. Tall buttercup is not quite as drought tolerant as bulbous buttercup, but it is equally unpalatable to livestock.

Bulbous buttercup growth habit and flowers

Bulbous buttercup bulb

Bulbous buttercup flower

Tall buttercup flower (note non-recurved sepals) and seed head

Tall buttercup growth habit

Tall buttercup flowers

Ranunculus repens L.   Creeping Buttercup

SYNONYMS: spotted buttercup, creeping crowfoot

LIFE FORM: **herbaceous perennial**; up to a foot (30 cm) tall

PLACE OF ORIGIN: Eurasia

VEGETATIVE CHARACTERISTICS: Creeping buttercup is typically a low-growing plant with leaves arranged in a basal rosette. The dark green leaves are often mottled with faint white markings; they are roughly 4 inches (10 cm) long and wide and consist of 3 distinct segments, the middle one of which is distinctly stalked. The margins of all the leaflets are irregularly lobed.

FLOWERS AND FRUIT: Creeping buttercup produces shiny, bright yellow insect-pollinated flowers from May through July that are about an inch (2.5 cm) across and have 5–7 petals; the sepals are not reflexed back against the stem and the flower stalk is distinctly grooved. The tiny, spherical fruits have a curved beak.

GERMINATION AND REGENERATION: The seeds germinate readily in moist soils; established plants produce runners (solons) that creep along the ground, rooting as they go, forming large, spreading patches.

HABITAT PREFERENCES: Creeping buttercup grows best in moist, humus-rich soil but also thrives in poorly drained sandy or gravelly soil; it seems to grow equally well in sun or shade. In its native habitat, creeping buttercup grows in grasslands and along stream banks.

ECOLOGICAL FUNCTION: Disturbance-adapted colonizer of moist ground.

RELATED SPECIES: **Lesser celandine (*Ranunculus ficaria* L.)** is an early-blooming, clump-forming European species that reproduces from fleshy tubers located both at the base of the plant and in the upper leaf axils. Its glossy, heart- or kidney-shaped leaves are up to 1.5 inches (4 cm) long and have wavy, toothed edges. Lesser celandine's solitary flowers are larger and showier than those of creeping buttercup, and it has been cultivated as an ornamental in gardens, from where it has managed to escape. The plant can become dominant in moist, sunny locations with rich soil but is also quite shade tolerant. Lesser celandine has been used in traditional European medicine to treat hemorrhoids.

Creeping buttercup in flower

Creeping buttercup foliage and flowers

Lesser celandine flowers

Lesser celandine foliage

Lesser celandine growth habit

Potentilla argentea L.   Silvery Cinquefoil

SYNONYM: silvery five-fingers

LIFE FORM: **herbaceous perennial**; up to 1.5 feet (45 cm) tall

PLACE OF ORIGIN: Eurasia

VEGETATIVE CHARACTERISTICS: Silvery cinquefoil can grow as an erect plant when left undisturbed or as a prostrate mat when mowed. The alternate, palmately compound leaves consist of 5–7 coarsely toothed leaflets about an inch (2.5 cm) long; their upper surface is glossy and their underside is covered with a dense mat of silvery white hairs (the source of the common name).

FLOWERS AND FRUIT: From late spring through the summer, silvery cinquefoil produces small yellow flowers that have 5 petals and are about 0.5 inch (1 cm) in diameter; they can either be insect- or self-pollinated. In May and June the plant can transform compacted lawns into a sea of yellow. The small, dry fruits (achenes) are clustered together on a hairy receptacle.

GERMINATION AND REGENERATION: The seeds germinate readily on dry, sunny sites and are capable of germinating after a long period of burial in the soil. Established plants produce new shoots from a deep-rooted perennial crown.

HABITAT PREFERENCES: Silvery cinquefoil grows best in full sun and dry, compacted soils with low fertility.

ENVIRONMENTAL FUNCTION: Disturbance-adapted colonizer of bare ground.

CULTURAL SIGNIFICANCE: In traditional medicine, the root is considered an astringent and has a variety of minor uses.

RELATED SPECIES: **Oldfield cinquefoil (*Potentilla simplex* Michx.)**, also known as "five fingers," is a low-growing native species with palmately compound leaves that have 3 or 5 coarsely toothed leaflets. Wiry red or green runners (stolons) grow out from established plants, strawberry style, to produce new plantlets. The foliage typically dies to the ground in winter but can stay green in mild climates. The plant produces bright yellow flowers singly on long stalks from May through June; they are 0.5–0.75 inch (1–2 cm) wide and have 5 broad petals. The small, dry fruits (achenes) are clustered together on a hairy receptacle. The plant grows best in sunny locations and is an indicator of sandy soils that are acidic, low in nutrients, or both.

Silver cinquefoil in flower is producing the yellow haze in this urban park

Silver
cinquefoil
foliage

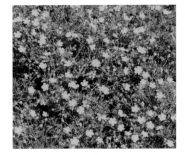

Silver cinquefoil flowers

Old-field cinquefoil habit and foliage; note the long runners

Old-field cinquefoil flower

Potentilla recta L.   Sulfur Cinquefoil

Synonyms: sulfur five-fingers, upright cinquefoil, rough-fruited cinquefoil

Life Form: herbaceous perennial; up to 30 inches (75 cm) tall

Place of Origin: Eurasia

Vegetative Characteristics: Sulfur cinquefoil stems are hairy and upright; the hairy, palmately compound leaves consist of 5–7 lance-shaped leaflets—up to 2 inches (5 cm) long—with coarsely toothed margins and a long petiole. The upper leaves are smaller and have only 3 leaflets and a short or no petiole. The plant produces a strong taproot.

Flowers and Fruit: Showy, pale yellow flowers are produced from June through August. The flowers—which are arranged in flat clusters at the ends of the branches—are about an inch (2.5 cm) in diameter and have 5 petals that are shallowly notched at the tip; they can either be insect- or self-pollinated. The dry fruits (achenes) are clustered together on a hairy receptacle.

Germination and Regeneration: The seeds germinate readily in sunny, dry sites and can survive burial in the soil for many years. Established plants produce new shoots from dormant buds located at the periphery of the old stem, eventually forming a circular cluster of plants.

Habitat Preferences: Sulfur cinquefoil grows well in sunny, dry locations. In the urban environment it is common in waste areas, dumps, unmowed highway banks, urban meadows, and roadside ditches.

Ecological Function: Disturbance-adapted colonizer of bare ground.

Cultural Significance: The pale yellow flowers are quite beautiful, but the plant has a tendency to spread and is considered an invasive species in the overgrazed pasture lands of the northern Rocky Mountains.

Related Species: Rough cinquefoil (*Potentilla norvegica* L.) is an annual or biennial species with stout, hairy stems up to 3 feet (1 m) tall. Its leaves have 3 leaflets that are hairy and green on both surfaces. Its small yellow flowers—less than 0.5 inch (1 cm) wide—have 5 petals that are shorter than the green sepals below them. Despite its species name, rough cinquefoil is native to both Europe and North America.

Sulfur cinquefoil growth habit

Lower leaves of sulfur cinquefoil

Close-up of sulfur cinquefoil flower

Sulfur cinquefoil flowers and developing fruits

Rough cinquefoil
in bloom

Galium mollugo L.   Smooth Bedstraw

SYNONYMS: wild madder, hedge bedstraw, false baby's breath, white bedstraw

LIFE FORM: **herbaceous perennial**; with stems up to 3 feet (1 m) long

PLACE OF ORIGIN: Eurasia

VEGETATIVE CHARACTERISTICS: Smooth bedstraw is a weakly upright or sprawling plant with square stems that are smooth to the touch. The narrow, lance-shaped leaves lack petioles and are about 0.5–1.5 inches (1–4 cm) long; 6–8 leaves are arranged in distinct whorls or layers at each node on the stem.

FLOWERS AND FRUIT: Conspicuous clusters of tiny white flowers with 4 petals are produced at the ends of the branches in late spring and summer, giving the plant a wispy appearance. Following insect-pollination, smooth bedstraw produces 2-lobed fruits that lack hooks on their surface.

GERMINATION AND REGENERATION: The seeds germinate readily in a variety of habitats; established plants produce yellow rhizomes and stolons that allow it to spread extensively.

HABITAT PREFERENCES: Smooth bedstraw grows best in shady or sunny, moist, nutrient-rich soil, but it can also grow in dry sandy sites. In the urban environment it is common along roadsides and pathways, minimally maintained lawn and ornamental planting beds, worn-out lawns and meadows, and the moist edges of wetlands.

ECOLOGICAL FUNCTIONS: Disturbance-adapted colonizer of bare ground; erosion control through its tenacious roots.

RELATED SPECIES: **Catchweed bedstraw** or **cleavers** (*Galium aparine* L.) is an annual species that appears to be native to North America, Europe, and Asia. Its square stems, leaves, and fruits are all covered with short, downward-pointing "hooks" that catch hold of clothing and give the plant its common name. It produces small white flowers in early summer and its prostrate or climbing stems can grow up to 3 feet (1 m) long. Extracts of its leaves have long been used as a "spring tonic" and a diuretic to treat kidney and bladder problems; the seeds, when roasted, have been used as a coffee substitute.

Field of smooth bedstraw in Vermont

Wispy smooth bedstraw flowers

Smooth bedstraw foliage

Smooth bedstraw in bloom

Catchweed bedstraw foliage

Catchweed bedstraw fruits annoyingly stick to clothing

Linaria vulgaris Mill.   Yellow Toadflax

SYNONYMS: butter and eggs, eggs and bacon, toadflax, Ransted-weed, wild snapdragon, Jacob's ladder, flaxweed, continental weed, patterns and clogs, pedlar's basket, fluellin, devil's ribbon, eggs and collops, buttered haycocks, etc.

LIFE FORM: herbaceous perennial; up to 3 feet (1 m) tall

PLACE OF ORIGIN: Europe

VEGETATIVE CHARACTERISTICS: The smooth, unbranched stems of yellow toadflax arise from underground stems to form large clumps. The narrow, linear leaves are alternate, pale gray-green, and 0.5–1.5 inches (1.5–4 cm) long; under stressful conditions the leaves can be so tightly packed together that the stems have a bottle-brush appearance.

FLOWERS AND FRUIT: Yellow toadflax produces compact clusters of pale yellow, snapdragon-like flowers at the ends of the stems from June through October. Individual flowers have 5 petals that are fused to form 2 lips; the lower lip has a conspicuous orange throat and an elongated spur filled with nectar, to attract insects. The flowers are an inch or so (2–3.5 cm) long and are pollinated by bees. The fruit is an egg-shaped capsule containing numerous tiny long-tongued seeds.

GERMINATION AND REGENERATION: Yellow toadflax reproduces from seed; the taproot and lateral roots of established plants give rise to adventitious shoots that can form large, often circular colonies. New shoots are also produced at the base of old stems and remain green through the winter.

HABITAT PREFERENCES: Yellow toadflax tolerates drought, full sun, and compacted soil. It is extremely common in the urban environment, where it can be found in minimally maintained public parks, vacant lots, rubble dumps, urban meadows, small pavement openings and cracks, rock outcrops and stone walls, unmowed highway banks, and railroad rights-of-way. In its native habitat yellow toadflax is found in grasslands, open forests, sand dunes, and gravelly areas.

ECOLOGICAL FUNCTIONS: Disturbance-adapted colonizer of bare ground; erosion control on slopes; tolerant of roadway salt and compacted soil.

CULTURAL SIGNIFICANCE: Yellow toadflax has a long history of medicinal use in Europe as an astringent, a purgative, and a diuretic to treat jaundice and liver problems. It arrived early in North America; Josselyn reported it growing in New England in 1672. The plant inspired John Bartram (1759) to publish what may be the first report of a "plant invasion" in North America: "ye stinking yellow linarya is ye most hurtfall plant in our pastures that can grow in our Northern climate. Neither the spade plow nor hoe can destroy it. When it is spread in a pasture every little fiber that is left will increase prodigiously. . . . It is now spread over great part of ye inhabited parts of Pennsylvania. It was first introduced a fine garden flower but never was a plant more heartily cursed by those that suffers by its incroachment."

RELATED SPECIES: Oldfield toadflax (*Nuttallanthus canadensis* (L.) D.A. Sutton [formerly *Linaria canadensis*]), a native of North America, grows to about 2 feet (0.6 m) tall and has thin, narrow leaves about an inch (2.5 cm) long. It produces numerous tiny blue flowers at the ends of tall, slender stalks from May through September. It is common in sunny, dry soil along the margins of roadsides and walkways—and is a tough plant despite its delicate appearance.

Yellow toadflax growth habit

Yellow toadflax in dry gravel

Yellow toadflax foliage

Yellow toadflax flowers

Oldfield toadflax growing by a roadside

Close-up of oldfield toadflax flowers

SYNONYMS: Aaron's rod, Jacob's staff, our lord's candle, flannel leaf, velvet dock, big taper, candlewick, great mullein, wooly mullein, lungwort, high taper, etc.

LIFE FORM: **evergreen biennial**; up to 6 feet (2 m) tall

PLACE OF ORIGIN: Eurasia

VEGETATIVE CHARACTERISTICS: During its first year mullein forms a flat rosette of woolly gray-green leaves that persist through the winter; in its second year the plant produces a tall, unbranched flower stalk. The alternately arranged leaves are oblong or lance shaped; those at the base of the plant can be up to a foot (30 cm) long, and they become progressively smaller up the stalk. The plant produces a thick, fleshy taproot.

FLOWERS AND FRUIT: The sulfur yellow flowers are produced from June through August on an unbranched terminal spike that can be up to 18 inches (50 cm) long; vigorous plants sometimes produce branched flower stalks. Individual flowers have 5 fused petals, are about an inch in diameter, and are pressed tightly against the stalk (sessile). They are either insect- or self-pollinated. The fruit capsules are arranged spirally around the flower stalk, and each contains several hundred tiny, dustlike seeds. A robust plant can produce well over 100,000 seeds.

GERMINATION AND REGENERATION: Mullein reproduces readily from seeds, which germinate in fall or spring, typically on freshly disturbed ground. They are dispersed only a short distance as the stalk moves back and forth in the wind. Mullein is the quintessential colonizer: it germinates from buried seed following soil disturbance, produces a large crop of seeds, and then disappears until the next round of disturbance unearths the seeds. In Europe, seeds buried for up to 500 years germinated when brought to the surface.

HABITAT PREFERENCES: Mullein grows best on dry, sandy or gravelly soil that has been recently disturbed, especially on steep slopes where little else is growing. In the urban environment it is common in vacant lots, rubble dumps, rock outcrops and stone walls, unmowed highway banks, and railroad rights-of-way.

ECOLOGICAL FUNCTION: Disturbance-adapted colonizer of bare areas.

CULTURAL SIGNIFICANCE: Dioscorides included mullein in his first-century herbal, *De Materia Medica*, and it has a long history in European folk medicine as a tea to treat chest colds, asthma, bronchitis, and kidney infections; and as a poultice to reduce swelling. Early settlers inserted the thick, fuzzy leaves into their shoes and clothing for extra insulation in the winter. Mullein was introduced into North America in the late 1600s or early 1700s, probably for medicinal purposes. In 1859, Darlington noted that "there is no surer evidence of a slovenly, negligent farmer, than to see his fields over-run with Mulleins. When neglected, the soil soon becomes so full of seeds that the young plants will be found springing up, in great number, for a long succession of years."

RELATED SPECIES: Moth mullein (*Verbascum blattaria* L.), another biennial with long-lived seeds from Eurasia, produces a rosette of nonhairy leaves with toothed margins in its first year and a tall flower stalk in its second. Its flowers are about an inch (2.5 cm) across and are white or yellow with a reddish purple center.

Common growing mullein on a bed of blacktop

Common mullein rosette at the end of its first year

Common mullein at the end of its second year

Close-up of common mullein in flower with spotted knapweed

Close-up of common mullein flowers

Veronica arvensis L.   Corn Speedwell

Synonyms: rock speedwell, wall speedwell

Life Form: winter annual; up to 6 inches (15 cm) tall

Place of Origin: Europe

Vegetative Characteristics: Corn speedwell can either be upright in its growth habit or form a mat with prostrate stems that radiate out from the base. The leaves are opposite, egg shaped, 0.25–0.50 inch (6–12 mm) long, and have rounded teeth on the margins and a distinct petiole. The leaves and stems are distinctly hairy. Leaves on the upright flowering stems are alternately arranged, crowded together, and much smaller and narrower than the ones lower down on the stem, and they lack a distinct petiole. The whole plant typically dies in summer.

Flowers and Fruit: Corn speedwell produces its blue flowers in mid-to-late spring on stalks that stand well up above the foliage. The flowers have very short pedicles, are about an eighth of an inch (3 mm) wide, and have 4 petals and 2 prominent stamens that protrude outward from the center; they can be either insect- or self-pollinated. The small, heart-shaped fruits that follow contain numerous tiny seeds.

Germination and Regeneration: The seeds germinate mainly in fall or spring, but will also sprout in summer if the weather is cool and moist. Stems root at the nodes where they touch the ground. The buried seeds of this species and most other *Veronicas* can remain viable for many years.

Habitat Preferences: Corn speedwell can grow in a wide range of soil conditions but does best in dry, sandy or gravelly soil in full sun. It is common in compacted soil in lawns and around the base of buildings, the margins of woodlands, ornamental planting beds, and small pavement openings and cracks.

Ecological Function: Disturbance-adapted colonizer of bare ground.

Cultural Significance: Speedwells have a long history of use in traditional European medicine as astringents to treat skin problems and as teas to treat coughs and to purify the blood.

Related Species: About a dozen species of speedwell grow in lawns and disturbed sites throughout the Northeast. **Purslane speedwell (*Veronica peregrina* L.)** is an annual that is similar to corn speedwell in growth habit but has unlobed, hairless, oval leaves and hairless stems; its flowers are white with short pedicels and about the same size as those of corn speedwell. **Thymeleaf speedwell (*Veronica serpyllifolia* L.)** is a perennial species that produces a dense, ground-hugging mat of dark green, egg-shaped leaves with short petioles. The flowers are white with blue highlights, about 0.25 inch (6 mm) wide, and are carried on stalks that stand up well above the mat of foliage. All three species of *Veronica* described here are native to Europe, and it is not unusual to find them growing together at the same site.

Close-up of corn speedwell flower and foliage

Corn speedwell growth habit near the end of its life span; plants are about 4 inches (10 cm) tall

Purslane speedwell growth habit

Purslane speedwell flowers and foliage

Close-up of thymeleaf speedwell flower

Thymeleaf speedwell growth habit

Solanum nigrum L.   Black Nightshade

SYNONYMS: deadly nightshade, poison berry, garden nightshade, common nightshade, European black nightshade, houndsberry

LIFE FORM: **summer annual** or **short-lived perennial**; up to 2 feet (70 cm) tall

PLACE OF ORIGIN: Eurasia

VEGETATIVE CHARACTERISTICS: Black nightshade is a low-branched plant with green stems that are smooth or slightly hairy, but not sticky to the touch. The alternate leaves are oval to egg shaped and have irregular, blunt teeth along their margins and a dull purplish underside. The leaves are 1–5 inches (2–12 cm) long by half as wide, and always seem to be covered with tiny holes created by insects that feed on them.

FLOWERS AND FRUIT: The white, star-shaped flowers, which are produced in late summer, have yellow anthers and are less than 0.5 inch (1 cm) wide; they are arranged in small, drooping clusters at intervals along the stem and can either be insect- or self-pollinated. The fruit is a round, glossy black berry containing numerous yellow or white seeds.

GERMINATION AND REGENERATION: Black nightshade fruits are consumed by birds, and the seeds germinate beneath their roosting areas as well as in various other habitats. Seeds can retain their viability in the soil for many years.

HABITAT PREFERENCES: Black nightshade grows in a wide variety of disturbed habitats but prefers sunny, moist soil. In the urban environment it is common in neglected ornamental landscapes, the margins of minimally maintained parks, vacant lots, rubble dumps, small pavement openings, sidewalk cracks, compacted soil along walkways, stone walls, and rock outcrops.

ECOLOGICAL FUNCTION: Disturbance-adapted colonizer of bare ground.

CULTURAL SIGNIFICANCE: Dioscorides included black nightshade in his first-century herbal, *De Materia Medica*. While the plant is somewhat toxic and narcotic in its effects, it has been used in traditional medicine to treat inflammations, skin disorders, and eye problems in Europe, India, Asia, and Africa. The stems and leaves are poisonous to livestock and people, as are the green fruits. The fully mature black fruits, however, are nontoxic and have reportedly been used to make pies and preserves. The plant was an early arrival in North America and is listed in Josselyn's *New-England's Rarities* (1672) under the category: "Of such plants as have sprung up since the English planted and kept cattle in New England."

RELATED SPECIES: **Eastern black nightshade (*Solanum ptycanthum* Dunal)** is a native North American species that is quite common in the East and is difficult to distinguish from European black nightshade. Indeed, even taxonomists cannot agree about the exact classification of the various species that make up the "*Solanum nigrum* complex."

Black nightshade growth habit

Purplish leaf undersides of black nightshade

Black nightshade flowers

Black nightshade foliage is often damaged by insects

Black nightshade fruits

Humulus japonicus Sieb. & Zucc.   Japanese Hops

LIFE FORM: herbaceous annual vine; up to 10 feet (3 m) long

PLACE OF ORIGIN: temperate eastern Asia

VEGETATIVE CHARACTERISTICS: The stems and leaves of Japanese hops have a rough texture due to the presence of hooked bristles that catch hold of anything they touch. The dull green leaf blades have 5–7 lobes and a heart-shaped base and are 2–5 inches (5–13 cm) long; on most leaves the petiole is longer than the blade. The plant has more of a sprawling than a climbing growth habit and can easily overwhelm adjacent shrubs.

FLOWERS AND FRUIT: Japanese hops produces wind-pollinated flowers in midsummer on separate male and female plants (dioecious). The male spikes are conspicuous and up to 10 inches (25 cm) long; female spikes are inconspicuous and only about an inch (2 cm) long. The sticky fruit clusters produced by female plants ripen in late summer.

GERMINATION AND REGENERATION: The seeds germinate readily in sunny, disturbed sites with moist soil.

HABITAT PREFERENCES: Japanese hops grows best on moist, rich soil in full sun. It is commonly found growing on compost and rubbish piles and river floodplains.

ECOLOGICAL FUNCTIONS: Erosion control; food and habitat for wildlife.

CULTURAL SIGNIFICANCE: Japanese hops was introduced into North America in the late nineteenth century but did not attract attention until the 1930s and 1940s, when it spread explosively throughout the East. It is much more vigorous and adaptable than European hops (*Humulus lupulus*), whose fruits are used to flavor beer. While several variegated varieties of Japanese hops are cultivated for ornamental purposes, none are used in beer making.

Japanese hops foliage

Japanese hops male inflorescence

Young shoot of Japanese hops with rough hairs on the stems and leaves

Japanese hops scrambling over a pile of rubble

Developing seed heads of Japanese hops

Urtica dioica L. var. *procera* (Muhl.) Wedd.   Stinging Nettle

SYNONYMS: *Urtica procera*, slender nettle

LIFE FORM: **herbaceous perennial**; up to 6 feet (2 m) tall

PLACE OF ORIGIN: eastern North America

VEGETATIVE CHARACTERISTICS: This erect, unbranched plant has square, grooved, bristly stems. The narrow, opposite, egg- to lance-shaped leaves are 2–6 inches (5–15 cm) long with a rounded base and coarsely serrated margins. The leaves are arranged in pairs that are perpendicular to one another. Their lower surface is covered by silica-tipped, hollow stinging hairs that release a skin irritant (formic acid) when touched.

FLOWERS AND FRUIT: Stinging nettle produces pendulous chains of greenish yellow, wind-pollinated flowers in the axils of the uppermost leaves from late spring through summer. Individual flowers are small and lack petals; male and female flowers are separate from one another but on the same plant (monoecious).

GERMINATION AND REGENERATION: The small seeds are dispersed passively as the plant blows in the wind; they germinate readily in moist soil. Established plants produce deeply rooted rhizomes that give rise to numerous new shoots and can form large colonies. Stinging nettles can be difficult to eliminate because of the tenacity of their rhizomes.

HABITAT PREFERENCES: Stinging nettle grows best in moist, nutrient-rich soil in either sun or shade. In the urban environment it is common along the margins freshwater wetlands, ponds, and streams; in moist soil at the edge of woodlands; and in roadside drainage ditches.

ECOLOGICAL FUNCTIONS: Disturbance-adapted colonizer of moist, bare soil; nutrient cycling in rich soil.

CULTURAL SIGNIFICANCE: Brushing against nettle leaves results in a painful, burning sting that is not soon forgotten. When cooked, nettles lose their sting and are safe to eat. In spring the young shoots can be collected and boiled to make tea or, when pureed, a nutritious soup. The stem fibers can be used to make paper and a durable canvas-like cloth; the roots and leaves produce strong dyes.

RELATED SPECIES: **European stinging nettle** (*Urtica dioica* L. var. *dioica*) is a smaller, weaker plant than its American cousin, and its broader leaves have a heart-shaped base. European stinging nettle is dioecious rather than monoecious, and the stinging hairs are located on both surfaces of the leaf blade. Dioscorides included stinging nettle in his first-century herbal, *De Materia Medica*, and indeed, every part of the plant seems to have some medicinal use. In the Middle Ages people deliberately applied nettles to their skin not only as a form of religious penance (urtication), but also as a treatment for rheumatism and the loss of muscle strength. Josselyn listed stinging nettle in *New-England's Rarities* (1672) under the heading: "Of such plants as have sprung up since the English planted and kept cattle in New England."

Stinging nettle growth habit

Young stinging nettle shoots

Stand of stinging nettles in spring

Stinging nettle flowers in late summer

Stinging nettle flowers and foliage

Viola sororia Willd.   Common Blue Violet

Synonyms: *Viola papilionacea,* meadow violet, hooded blue violet, dooryard violet

Life Form: herbaceous perennial; up to 8 inches (20 cm) tall

Place of Origin: eastern North America

Vegetative Characteristics: Common blue violet is a low-growing plant with glossy, heart-shaped leaves—up to 4 inches (10 cm) wide—arising from a basal crown. The leaf blades are smooth, highly variable in size, and have small, rounded teeth along their margins; the petioles are roughly twice as long as the leaf blades.

Flowers and Fruit: Common blue violet is well known for the flowers it produces from April through June; they come in a variety of shades of purple, blue, violet, and white—blue with a white throat being the most common. The flowers are about 1 inch (2.5 cm) wide, have 5 petals, and are carried on stalks that are about the same height as the leaves. In addition to the conspicuous, insect-pollinated flowers it produces in spring, common blue violet also produces inconspicuous white flowers at or just below ground level later in the season. These self-pollinating, cleistogamous (hidden) flowers never open and produce far more seeds than the aerial flowers.

Germination and Regeneration: The common blue violet reproduces readily from seed; established plants produce short rhizomes that give rise to new plants in the spring. Under the right conditions they can come to dominate large patches of lawn.

Habitat Preferences: Common blue violet grows best in cool, moist, shady soil, but it can also tolerate sites in full sun with dry soil. In the urban environment it is common in minimally maintained lawns (it is tolerant of mowing), damp woodland thickets, roadside drainage ditches, and small holes in pavement.

Ecological Function: Disturbance-adapted colonizer of bare ground.

Cultural Significance: Native Americans used an infusion of common blue violet to treat a variety of minor ailments. The young leaves are edible in spring, as are the flowers, which can be eaten raw or candied.

Common blue
violet growth
habit

Underground flowers of common
blue violet

Common blue violet foliage

White-flowered form of common
blue violet

Typical common blue violet flower

Commelina communis L.   Asiatic Dayflower

SYNONYM: common dayflower

LIFE FORM: **summer annual**; up to 2 feet (60 cm) tall and 2.5 feet (75 cm) wide

PLACE OF ORIGIN: Asia

VEGETATIVE CHARACTERISTICS: The smooth, green stems of Asiatic dayflower have distinctly swollen nodes or joints; their growth is mainly horizontal, but with upright shoot tips. The shiny, somewhat succulent leaves are lance-shaped to oblong and 2–5 inches (5–12 cm) long by 0.5–1.5 inches (1.5–4 cm) wide; they have parallel veins, and the sheaths at the base of the leaf blade clasp the stem. The root system is mostly adventitious and shallow-growing. The whole plant dies with the first frost.

FLOWERS AND FRUIT: The strikingly beautiful flowers emerge from a folded, leafy bract; they are 0.5–1 inch (1–2.5 cm) wide and have 2 prominent blue petals on top and 1 smaller white petal below. The bisexual flowers are open for only a day (hence the name) and are insect-pollinated; they are produced throughout the summer and into the fall. The 3 cross-shaped yellow structures at the center of the flower are sterile anthers, which make the flowers more attractive to potential pollinators.

GERMINATION AND REGENERATION: The seeds germinate readily; the stems root at the nodes where they touch the ground, which allows the plant to spread widely.

HABITAT PREFERENCES: Asiatic dayflower grows best in moist, shady areas but, because of its succulent nature, also tolerates shady, dry conditions. It is common in minimally maintained public parks; the margins of freshwater wetlands, ponds, and streams; the understory of horticultural plantings; the rain shadow of buildings and fences; and small pavement openings. Because it is intolerant of mowing, it is not typically found in lawns.

ECOLOGICAL FUNCTION: Disturbance-adapted colonizer of bare ground.

CULTURAL SIGNIFICANCE: A tea made from the leaves is used to treat sore throat in China. Recent research from China has shown that Asiatic dayflower growing on toxic mine spoils bioaccumulated copper and other heavy metals to high levels, an observation that suggests it might be useful for phytoremediation.

Asiatic dayflower set off by colored mulch

Asiatic dayflower growth habit

Asiatic dayflower foliage

Asiatic dayflower as an accidental ornamental

Close-up of Asiatic dayflower flower

Cyperus esculentus L.   Yellow Nutsedge

SYNONYMS: yellow nutgrass, nutsedge, chufa, galingale, umbrella sedge, rushnut

LIFE FORM: **herbaceous perennial**; up to 2 feet (60 cm) tall

PLACE OF ORIGIN: Europe, Asia, and North America

VEGETATIVE CHARACTERISTICS: The shiny, grasslike leaves of yellow nutsedge are less than 0.3 (8 mm) wide and have a distinctive yellowish green tint that makes them stand out in the landscape. The upright flowering stems are unbranched and triangular (all botany students learn that "sedges have edges"). The plant dies to the ground with frost.

FLOWERS AND FRUIT: The clusters of shiny, yellowish brown flower spikelets, 0.5–1.25 inches (1–3 cm) long, are arranged horizontally and look like miniature bottle-brushes; they are produced near the top of the flower stalk and are subtended by thin, leafy bracts. Yellow nutsedge flowers are wind-pollinated.

GERMINATION AND REGENERATION: The seeds germinate on moist, bare soil; established plants send out long, scaly rhizomes that terminate with succulent tubers ("chufas"). These grow outward the following spring and are responsible for the plant's ability to spread rapidly.

HABITAT PREFERENCES: Yellow nutsedge grows in a variety of disturbed habitats but prefers sunny, moist sites and nutrient-rich soil. It is common in landscape planting beds; poorly drained turf areas; and on the margins of freshwater wetlands, ponds, and streams.

ECOLOGICAL FUNCTIONS: Disturbance-adapted colonizer of moist, bare ground; food for wildlife.

CULTURAL SIGNIFICANCE: The underground tubers can be eaten raw, cooked, or dried and have a history of use dating back to the ancient Egyptians. Cultivated varieties are grown in Africa and China.

RELATED SPECIES: **False nutsedge (*Cyperus strigosus* L.)** is native to North America and resembles yellow nutsedge except that its foliage lacks the yellowish cast; also it does not reproduce from rhizomes or tubers, so it does not spread as rapidly. It prefers moist, sunny sites and its seeds are capable of germinating after long periods of burial in the soil. Established plants expand slowly through the production of short, thick rhizomes.

Yellow nutsedge growth habit

Yellow nutsedge reproduces prolifically from rhizomes

Yellow nutsedge in the median strip

Yellow nutsedge flowers

Developing seeds of false nutsedge

Iris pseudacorus L.   Yellow Flag Iris

Synonyms: water flag, yellow iris, pale yellow iris, Jacob's sword, daggers, fleur-de-lis (-lys), flower-de-luce

Life Form: herbaceous perennial; up to 5 feet (1.4 m) tall

Place of Origin: Europe, western Asia, and North Africa

Vegetative Characteristics: Yellow flag iris is a robust, clump-forming marsh plant with stiff, erect, sword-shaped leaves that are about 3 feet (1 m) long, an inch (2.5 cm) wide, and have a distinctive bluish green cast. As the season advances, the leaves get beaten down by the weather and become prostrate.

Flowers and Fruit: Clusters of showy, bright yellow to cream-colored flowers, 2.5–3 inches (7–9 cm) wide, are produced on tall, branched stalks in June. Following pollination by bees, cylindrical green fruits develop—up to 3 inches (6 cm) long—each containing 3 stacks of flattened, D-shaped seeds.

Germination and Regeneration: Seeds germinate readily in sunny, moist soils; established plants spread vigorously from thick, branching rhizomes that can form large, dense mats of vegetation.

Habitat Preferences: Yellow flag iris grows well in a wide variety of freshwater and brackish wetlands, including the margins of ponds and streams; standing water up to 9 inches (23 cm) deep; and poorly drained, acidic soils. It grows well in either full sun or light shade and can tolerate periods of drought.

Ecological Functions: Stream and river bank stabilization; nutrient absorption in wetlands.

Cultural Significance: The flower is the inspiration of the fleur-de-lis, the heraldic emblem of the kings of France. Yellow flag iris also has traditional medicinal uses. Dioscorides included it in his first-century herbal, *De Materia Medica*, and in traditional European medicine the rhizome was used as a powerful cathartic as well as for a number of other less dramatic purposes, including as a cosmetic to remove bruises. Because of its high tannic acid content the root was also used for tanning leather. In recent times yellow flag iris has been planted in water treatment wetlands for its capacity to absorb nutrients and heavy metals (phytoremediation). The plant was introduced into North America in the mid-1800s, probably for ornamental purposes, and has spread spontaneously into eastern wetlands. Many states list it as an invasive species.

Yellow flag iris
growth habit

Yellow flag iris emerging in early spring

Yellow flag iris in early summer

Yellow flag iris in late fall

Yellow flag iris flower

Juncus tenuis Willd.　Path Rush

Synonyms: slender rush, wire grass, poverty rush

Life Form: evergreen perennial; up to 1 foot (0.3 m) tall

Place of Origin: Eurasia and North America

Vegetative Characteristics: Path rush is a clump-forming, grass-like plant with tough, round, hollow, dark green stems. The scraggly basal leaves, which are about half the height of the flowering stems, are flat with inwardly rolled edges and persist through the winter. The stems are extremely flexible and spring back after quickly being stepped on or driven over. The shallow, fibrous root system is remarkably tenacious.

Flowers and Fruit: Path rush produces clusters of inconspicuous greenish brown flowers near the top of the stiff stalks in late spring and early summer. Individual flowers, which have 3 petals and 3 long sepals, are subtended by long, leafy bracts. Following wind-pollination, small fruit capsules develop and split into 3 parts at maturity, releasing hundreds of orange-brown seeds that become sticky when wet.

Germination and Regeneration: The sticky seeds are transported by animals and people (as well as tire treads), and germinate readily on sunny sites with poor soil; established plants increase in size by means of short rhizomes. The seeds are capable of germinating after long periods of burial in the soil.

Habitat Preferences: As the common name suggests, this species thrives in compacted, drought-prone soil and is common along heavily traveled walkways as well as in pavement cracks with pedestrian or vehicular traffic. It also grows in heavy, wet soils and in roads paved with gravel.

Ecological Function: Disturbance-adapted colonizer of bare ground.

Cultural Significance: Path rush is one of the few plants treated in this book for which humans have found few uses. It was inadvertently introduced into Europe in the late 1700s and has become established on that continent in compacted soils along roadways.

Path rush growing in a roadway crack

Path rush growth habit

Path rush in the fall

Path rush in flower

Developing fruits of path rush

Allium vineale L.   Wild Garlic

Synonyms: field garlic, scallions, wild onion, crow garlic

Life Form: herbaceous perennial; up to 1–3 feet (30–90 cm) tall

Place of Origin: Europe, western Asia, and North Africa

Vegetative Characteristics: Wild garlic produces slender, hollow, grasslike leaves that smell strongly of onions or garlic when crushed. The leaves, which are up to 8 inches (20 cm) long, arise from a small basal bulb and are circular in cross section. Wild garlic is a cool-season plant that grows vigorously in spring and early summer, dies back in July, and reappears in fall.

Flowers and Fruit: Wild garlic produces round clusters of flowers and tiny bulblets at the top of sturdy green stalks in May or June. The aerial bulblets, which typically outnumber the flowers, grow to the size of a rice grain and sprout long, spidery leaves while still on the upright stalk. Wild garlic rarely—if ever—produces seeds in the Northeast.

Germination and Regeneration: As the viviparous bulblets expand, the flower stalk often bends over and touches the ground, thereby allowing them to take root and establish an independent existence. The mother plant regenerates in the spring from a perennial cluster of underground bulbs which are covered with a papery sheath.

Habitat Preferences: Wild garlic grows best in moist, nutrient-rich soils in sun or shade, but it can tolerate a wide range of habitats. It is common in minimally maintained public parks; disturbed woodlands; wet meadows; and the margins of freshwater wetlands, ponds, and streams.

Ecological Function: Disturbance-adapted colonizer of bare ground.

Cultural Significance: John Bartram (1759) described several reasons for not being fond of wild garlic: "Crow garlic—this is greatly loved by ye horses cows & sheep & very wholesome early pasture for them yet our people generally hates it because it makes ye milk butter cheese & indeed the flesh of those cattle that feeds much upon it taste so strong that we can hardly eat of it but for horses & young cattle it doth very well but our millers cant abide it amongst corn it clogs up their miss so that it is impossible to make good flower." The bulbs can be eaten in the spring.

Wild garlic in
early spring

Wild garlic in its typical urban niche

Wild garlic in the understory of an
Ailanthus grove

Wild garlic foliage grows back in the fall

Wild garlic bulblets sprouting
on top of the flower stalk

Smilax rotundifolia L.   Roundleaf Greenbrier

SYNONYMS: catbrier, bullbrier, hellfetter, blasphemy vine, horsebrier

LIFE FORM: perennial vine; with stems up to 15 feet (5 m) long

PLACE OF ORIGIN: eastern North America

VEGETATIVE CHARACTERISTICS: Roundleaf greenbrier is a low-climbing vine whose slender green stems are studded with stiff, sharp thorns that make the plant extremely unpleasant to encounter—almost like running into barbed wire. The glossy leaves are alternate, simple, rounded to heart shaped, and roughly 2–5 inches (5–13 cm) long and wide; they have smooth margins and distinctly parallel veins. The plant climbs by means of the tendrils that terminate the leaf stipules. Plants typically shed their leaves in late fall; the leafless stems maintain their olive green coloration through the winter.

FLOWERS AND FRUIT: Roundleaf greenbrier is a dioecious species with separate male and female (seed-producing) individuals. Inconspicuous greenish flowers are produced in small axillary clusters from April through August and are pollinated by a variety of insects; bluish black fruits develop in September.

GERMINATION AND REGENERATION: The fruits are dispersed by birds, and the seeds germinate in a variety of habitats. Established plants produce new stems from deep underground rhizomes, leading to the formation of dense, impenetrable thickets.

HABITAT PREFERENCES: Roundleaf greenbrier grows best on moist, sunny sites but also tolerates dry, shady conditions. It is common along roadsides; at the edges of thickets; and in the understory of low, moist woods.

ECOLOGICAL FUNCTIONS: Food and habitat for wildlife; tolerant of roadway salt and compacted soil; stream and river bank stabilization.

CULTURAL SIGNIFICANCE: The young shoots and leaves are edible in spring, and the roots produce a gelatin that can be used to thicken soups.

Roundleaf greenbriar leaf and spines

Thicket of roundleaf greenbriar stems in winter

Roundleaf greenbriar on a chain-link fence

Roundleaf greenbriar tendrils and sharp spines

Roundleaf greenbriar fruits

Bromus tectorum L. Downy Brome

Synonyms: cheatgrass, downy chess, early chess, thatch grass, gas station grass, drooping brome

Life Form: winter annual; up to 2 feet (60 cm) tall

Place of Origin: Eurasia and North Africa

Vegetative Characteristics: All parts of the downy brome plant are covered with soft, silky hairs that give it a "fuzzy" appearance. The young leaves are twisted as they emerge, making it look like they are spiraling upward. The flat, hairy leaf blades can be up to 8 inches (20 cm) long. The fibrous root system is extremely efficient at absorbing soil moisture, especially where very little is available.

Flowers and Fruit: Downy brome produces a distinctive drooping inflorescence from early spring through midsummer depending on the time of seed germination; the spikelets, which consist of 3–8 florets with prominent awns, are 0.5–1 inch (1–2.5 cm) long. Following wind-pollination the flower heads develop into soft, shiny, purplish seed heads that turn tawny brown at maturity.

Germination and Regeneration: The seeds germinate in the fall or spring, depending on local conditions; seeds that germinate in the fall are able to grow roots as long as the soil temperatures are above freezing and will resume growing in early spring.

Habitat Preferences: This drought-tolerant species flourishes in sandy or compacted soils throughout the urban environment. It is common in pavement cracks and openings, minimally maintained landscape plantings, highway banks and median strips, and along railroad tracks.

Ecological Functions: Disturbance-adapted colonizer of bare ground; soil building on degraded land.

Cultural Significance: The species name, *tectorum*, means "of roofs" in Latin and indicates that this plant was traditionally used in Europe for thatching. Downy brome was introduced into North America in the mid-1800s and has become a major problem in the rangelands of the West, where it displaces native grasses and promotes fire. It is an important spring forage grass for free-range livestock.

Related Species: Smooth or **Hungarian brome** (*Bromus inermis* **Leyss.**) is a cool-season perennial that was introduced in the 1880s as a forage plant for livestock and escaped to become a common roadside species. It gained prominence during the 1930s because of its drought tolerance. Smooth brome grows up to 4 feet (1.3 m) tall and produces a nodding, purple-tinged inflorescence in late spring or early summer. At maturity the seed heads droop conspicuously from the weight of the large, single-seeded fruits.

Downy brome
in its typical
urban niche

Downy brome in the city

Downy brome colonizing a pile of asphalt

Developing
seeds of
downy
brome

Mature
seed heads
of downy
brome

Smooth brome seed heads

Dactylis glomerata L.   Orchardgrass

Synonym: cocksfoot

Life Form: **evergreen perennial**; up to 4 feet (1.3 m) tall

Place of Origin: Europe

Vegetative Characteristics: Orchardgrass is a tall, clump-forming plant that stays green through the winter. The bluish-green leaf blades are up to 12 inches (30 cm) long and distinctly creased in the middle both as they emerge from the bud and when fully expanded. The densely clustered stems arise from perennial crowns and can form large clumps, or tussocks. The roots are very tenacious and can be pulled up only when the soil is wet.

Flowers and Fruit: Orchardgrass produces wind-pollinated flowers from spring through midsummer in spikelets that are crowded in 1-sided clusters at the end of stiff, upright branches. The clumped, fluffy appearance of the flowers and the seed heads makes this grass easy to identify in summer and fall.

Germination and Regeneration: Seeds germinate readily on sunny, bare soil and can germinate after being buried for many years. Established plants enlarge slowly through the growth of short rhizomes, often forming circular patches with a dead patch at their center.

Habitat Preferences: Orchardgrass is an adaptable species tolerant of drought, shade, and low soil fertility. It is extremely common in minimally maintained public parks, unmowed highway banks and median strips, vacant lots and abandoned property, and along sunny woodland edges.

Ecological Functions: Tolerant of roadway salt and compacted soil; erosion control on slopes; food for wildlife.

Cultural Significance: Orchardgrass was originally introduced from Europe as a forage crop for domestic animals and is still a major component of many hayfields in the East. Darlington (1759) explains the origin of its common name: "This grass also possesses the additional advantage of thriving well in the shade of trees, and answers a very good purpose in orchards."

Leaf collar of
orchardgrass

Orchardgrass
growth habit

Orchardgrass
dominates the
vast urban
meadows of
Detroit

Orchardgrass in early spring

Orchardgrass at maturity

Dichanthelium clandestinum (L.) Gould   Deer-tongue Grass

Synonyms: *Panicum clandestinum,* riverside panic grass

Life Form: **evergreen perennial**; up to 3 feet (1 m) tall

Place of Origin: eastern North America

Vegetative Characteristics: The upright to sprawling stems of this native grass have broad, stiff, lance-shaped leaves that are 3–4 inches (7–10 cm) long by an inch (2.5 cm) wide and are the source of its common name. The base of the leaf clasps the stem, and the sheath is distinctly hairy. The leaves turn brown in the fall and persist on the stems all winter, making this plant easy to identify. At first glance the plant looks like a small bamboo.

Flowers and Fruit: Deer-tongue grass produces wind-pollinated flowers from May through September. Flower stalks produced early in the season usually emerge completely, while those produced later in the season typically fail to emerge fully from their surrounding sheaths (hence the species name, *clandestinum*).

Germination and Regeneration: The seeds germinate in sun or shade and are capable of germinating after long periods of burial. Established plants increase in size through the growth of rhizomes and can form large clumps.

Habitat Preferences: Deer-tongue grass seems to grow best in moist, partly shaded conditions but is also found in dry, sunny locations. It is common along roadside drainage ditches; in the shady margins of wetlands, streams, and ponds; and in the compacted soil of vacant lots.

Ecological Functions: Disturbance-adapted colonizer of bare ground; food and habitat for wildlife; erosion control on slopes.

Deer-tongue grass growing along a heavily salted road

Deer-tongue grass foliage

Deer-tongue grass in winter

Leaf collar of deer-tongue grass

The inflorescence of deer-tongue grass often does not fully emerge from the leaf axil

Digitaria ischaemum (Schreb.) Schreb. ex Muhl.   Smooth Crabgrass

SYNONYMS: small crabgrass, fingergrass, twitchgrass

LIFE FORM: **summer annual**; up to 1 foot (30 cm) tall

PLACE OF ORIGIN: Europe

VEGETATIVE CHARACTERISTICS: Smooth crabgrass appears in early summer and grows rampantly until killed by frost. It typically has a prostrate or spreading growth habit and does not root at the nodes; the leaf blades and sheaths are smooth, as are the stems, which often develop a reddish coloration when stressed. The root system of established plants is extremely tenacious.

FLOWERS AND FRUIT: The skinny, wind-pollinated flower heads are composed of 2–6 finger-like spikes clustered at the top of a slender stem. The spikes are green when they emerge in late summer, then change to purple, and finally turn tan as the seeds mature.

GERMINATION AND REGENERATION: The seeds germinate readily in late spring and early summer and can remain viable in the soil for several years.

HABITAT PREFERENCES: Smooth crabgrass is remarkably drought tolerant and is common in trampled lawns in minimally maintained public parks and residential landscapes, unmowed highway banks and median strips, small pavement openings, and sidewalk cracks. In the urban environment smooth grass seems to be more abundant than hairy crabgrass (see below).

ECOLOGICAL FUNCTIONS: Disturbance-adapted colonizer of bare ground; tolerant of contaminated and compacted soil.

CULTURAL SIGNIFICANCE: Smooth crabgrass was once used as a forage grass in the south where cool-season grasses did not grow well. It often dominates patchy lawns, especially those that have been damaged by grubs or drought, and is typically the only green plant remaining in August.

RELATED SPECIES: **Large** or **hairy crabgrass** (*Digitaria sanguinalis* (L.) Scop.) is common in a variety of sunny disturbed habitats, including lawns, ornamental planting beds, pavement cracks, and median strips. It is more robust and upright than smooth crabgrass, and its stems root at the nodes. The leaves and stems are densely covered with fine hairs, and the inflorescence consists of 4–10 finger-like spikes clustered at the top of the stem. It is not as common in lawns as smooth crabgrass because it is taller and less tolerant of mowing. In Europe the seeds of large crabgrass have been used for food. **Goosegrass** (*Eleusine indica* (L.) Gaertn.) is a robust summer annual that resembles crabgrass but has flattened stems that radiate out from a central point and are silvery white at the base. If allowed to grow freely goosegrass can become 2–3 feet (60–90 cm) tall, but when mowed it forms a ground-hugging mat. Its flowers are produced in midsummer in jagged, finger-like clusters; its dry seed heads persist through the winter.

Smooth crabgrass in a typical urban niche

Smooth crabgrass growing along a roadway

Smooth crabgrass growth habit

Smooth crabgrass tolerates foot traffic

Hairy crabgrass growth habit

Hairy crabgrass seedling

Echinochloa crus-galli (L.) Beauv.   Barnyardgrass

SYNONYMS: cockspur grass, billion dollar grass, Japanese millet, barn-grass

LIFE FORM: summer annual; up to 5 feet (1.5 m) tall

PLACE OF ORIGIN: Eurasia

VEGETATIVE CHARACTERISTICS: Barnyardgrass is a vigorous, upright grower that produces multiple stems with adventitious prop roots at their base. The smooth leaf blades have a distinct light green to white midrib and can be 4–8 inches (10–20 cm) long and 0.2–0.8 inches (5–20 mm) wide; the sheaths are smooth, flattened, and reddish purple. The shallow root system makes the plant relatively easy to pull up.

FLOWERS AND FRUIT: The erect, terminal flower panicles are greenish red, coarsely branched, and covered with bristles. They are wind-pollinated and produced throughout the summer; the seed heads range from dark green to purple at maturity.

GERMINATION AND REGENERATION: The seeds germinate readily once the temperatures warm up in late spring.

HABITAT PREFERENCES: Barnyardgrass grows best in rich soil and full sun but can also tolerate shady, dry conditions. It is common in waste dumps throughout urban areas, minimally maintained landscape plantings, roadsides, and moist drainage ditches.

ECOLOGICAL FUNCTION: Disturbance-adapted colonizer of bare ground.

CULTURAL SIGNIFICANCE: Darlington (1759) explained that the common name of this species derives from the fact that "it is apt to abound along the drains of crude liquid flowing from barn-yards—and in spots which are usually designated as 'wet and sour.'"

Barnyardgrass growth habit

Leaf collar of
barnyardgrass

Barnyardgrass
foliage

Barnyardgrass
inflorescence

Barnyardgrass
mature seed

Elymus repens (L.) Gould   Quackgrass

Synonyms: *Agropyron repens, Elytrigia repens,* couch grass, quick grass, scratch grass, twitch grass, quake grass, scotch grass, creeping wheat

Life Form: evergreen perennial; up to 4 feet (1.4 m) tall

Place of Origin: Eurasia and North America

Vegetative Characteristics: Quackgrass is a "running" grass with stems (culms) emerging at some distance from one another. The bluish green leaves are rolled in the bud as they emerge and have sharp-pointed tips; the leaf blades can be up to a foot (30 cm) long. The crushed foliage has the distinctive smell of fresh wheat. The plant is evergreen and can establish large colonies that are difficult to eradicate. The name "quack grass" is a corruption of "quick grass," which refers to the fact that it greens up rapidly in spring.

Flowers and Fruit: Quackgrass produces a long, thin inflorescence with wind-pollinated flower clusters (spikelets) arranged alternately along the axis. The seed head is a long, narrow spike.

Germination and Regeneration: Seeds germinate readily on open ground; established plants spread rapidly by means of sharp-pointed white rhizomes that can produce both roots and shoots at the regularly spaced nodes.

Habitat Preferences: Quackgrass is common in minimally maintained landscape areas with good soil and full sun. It is also found in abandoned grasslands and meadows, small pavement openings, roadsides and median strips, and along railroad tracks.

Ecological Functions: Disturbance-adapted colonizer of bare ground; tolerant of roadway salt and compacted soil; erosion control on slopes.

Cultural Significance: Quackgrass was commonly planted in the nineteenth century as a forage crop valued for its ability to green up quickly and produce sweet hay. Eventually its aggressive, spreading habit became apparent and the plant was redefined as a weed that needed to be controlled. Rhizomes collected in the spring were used in European traditional medicine as a diuretic to treat a variety of urinary problems and to purify the blood, and they have also been used to make bread in times of famine.

Quackgrass growth habit

Quackgrass growing along a sidewalk

Quackgrass in its urban niche

Quackgrass in
bloom

Quackgrass
mature seed

Eragrostis pectinacea (Michx.) Nees. ex Steud.   Tufted Lovegrass

Synonym: Carolina lovegrass

Life Form: summer annual; 2–24 inches (5–60 cm) tall

Place of Origin: North America, except Alaska

Vegetative Characteristics: In good soil tufted lovegrass forms loose, open clumps up to 2 feet (60 cm) tall; in urban sidewalk cracks the clumps are denser and usually less than half that height. This sprawling grass produces thin, smooth stems; narrow leaves; and thin, flat flower stalks. New stems arise from the base and often lie flat on the ground but do not form adventitious roots. When in flower the whole plant has a shiny appearance and glistens in the sun.

Flowers and Fruit: Tufted lovegrass produces minute flowers on short spikelets that are arranged in a loose terminal inflorescence. Wind-pollinated flowers are produced from July through September, typically within a month or so of germination, followed by wispy seed heads.

Germination and Regeneration: The seeds germinate readily in sunny, moist or dry locations in late spring or early summer once the soil starts to warm up.

Habitat Preferences: This inconspicuous grass is very common in the urban environment, especially in sidewalk and pavement cracks. It also grows in a variety of disturbed sites in full sun, including vacant lots, rock outcrops, minimally maintained ornamental plantings, and highway banks and median strips.

Ecological Function: Disturbance-adapted colonizer of bare ground.

Related Species: Purple lovegrass (*Eragrostis spectabilis* (Pursh) Steud.) is another native species that is common along infrequently mowed highway edges that experience heavy salt applications. It grows to about 18 inches (45 cm) tall and, if not mowed, produces beautiful, dull red inflorescences that are extremely conspicuous in August and September.

Tufted lovegrass
in a sidewalk
crack

Tufted lovegrass growing only on the short
sides of bricks

Tufted lovegrass (with purslane) growing
between a curb and the street

Purple lovegrass dominating a highway median
strip in fall

Purple lovegrass in seed

Lolium arundinacea (Schreb.) S.J. Darbyshire   Tall Fescue

SYNONYMS: *Festuca arundinacea, Festuca elatior,* coarse or meadow fescue, tall ryegrass, Kentucky fescue

LIFE FORM: **evergreen perennial**; up to 2 feet (60 cm) tall

PLACE OF ORIGIN: Europe

VEGETATIVE CHARACTERISTICS: Tall fescue is easily recognized in the landscape by its broad, dark green leaves, whose glossy upper surfaces glisten in bright sunshine. The leaves are much coarser than most other lawn grasses and remain conspicuously green through the winter. As they emerge in spring the leaves are rolled (as opposed to folded) in the bud and lack a prominent midrib; on robust plants, the leaf blades can be up to 2 feet (60 cm) long by half an inch (12 mm) wide. The fibrous root system is deep and extremely tenacious.

FLOWERS AND FRUIT: Tall fescue produces relatively thin, wind-pollinated flower stalks in spring and early summer that grow to about 2 feet (60 cm) tall; they persist in a dry state into the fall.

GERMINATION AND REGENERATION: Tall fescue germinates readily from seed; established plants are long-lived and expand in size by producing new shoots (tillers).

HABITAT PREFERENCES: This species is highly tolerant of drought and low-fertility soils and grows well in either sun or shade; it is common on steep grassy slopes, minimally maintained turf areas, and along the compacted margins of paths and walkways.

ECOLOGICAL FUNCTIONS: Tolerant of roadway salt and compacted soil; erosion control on slopes. The leaves of this species contain an endophytic, which increases its tolerance of environmental stress.

CULTURAL SIGNIFICANCE: Tall fescue was introduced from Europe in the late 1800s as a pasture grass. Because of its toughness, it has become the most commonly planted grass for football fields and golf courses—covering literally tens of millions of acres across North America. It is also widely planted for erosion control on steep, difficult sites and is a common ingredient in drought-tolerant lawn seed mixes.

Related Species: **Red fescue (*Festuca rubra* L.)** is a clump-forming grass with wiry leaves that are much smaller and finer than those of tall fescue. It is native to Europe and is widely used in drought-tolerant lawn seed mixes. It naturalizes in semishaded, dry sites at the edge of emergent woodlands. Including its flower stalks, it grows to be about 2 feet (60 cm) tall.

Tall fescue foliage

Tall fescue holding a bank in winter

Tall fescue planted for erosion control

Tall fescue in flower

Widely spaced clumps of red fescue

Red fescue in flower

Muhlenbergia schreberi J.F. Gmel.   Nimblewill

Synonyms: wiregrass, drop-seed

Life Form: herbaceous perennial; up to 2 feet (60 cm) tall

Place of Origin: North America

Vegetative Characteristics: Nimblewill produces numerous slender, wiry stems that branch freely from the base and sprawl across the ground, eventually forming large clumps. The alternate gray-green leaves are narrow, sharp pointed, and only about 2 inches (5 cm) long. The prostrate stems typically form shallow-growing, adventitious roots at their lower nodes.

Flowers and Fruit: Nimblewill is one of the last grasses to bloom, usually in September and October. The slender inflorescences are 2–8 inches (5–20 cm) long and produced at the ends of both the terminal and the side branches; they have a somewhat wispy appearance, often with a slight purplish tint, and are wind-pollinated.

Germination and Regeneration: The seeds germinate readily; once the plant is established, the horizontal stems root at the nodes and can form large clumps.

Habitat Preferences: Nimblewill tolerates a wide range of moisture and light conditions from full sun to dense shade. In shady, dry sites (e.g., the rain shadow of buildings) it is quite spindly; on moist, rich soil it can be very robust. It tolerates compaction and is a common component of neglected lawns and grasslands where mowing encourages its spread.

Ecological Function: Disturbance-adapted colonizer of bare ground.

Related Species: Bermudagrass (*Cynodon dactylon* (L.) Pers.) is a warm-season perennial that spreads by both underground and aboveground runners that root at their jointed nodes. A native of Africa, it is typically considered a southern species, but cold-hardy varieties have been found growing spontaneously as far north as in New England. Bermudagrass can tolerate a wide range of ecological conditions, including drought, physical disturbance, salt, and both high and low soil pH. The 3–7 flower spikes are produced at the top of the stems from July through October; these skinny "fingers" superficially resemble those of crabgrass (*Digitaria* spp.), but they are shorter and thicker and radiate out from a central point. Bermudagrass is an important forage plant in parts of the southern United States.

Nimblewill in spring

Nimblewill along a sidewalk

Nimblewill growth habit

Nimblewill in flower in late summer

Panicum dichotomiflorum Michx.   Fall Panicum

Synonym: smooth witchgrass

Life Form: summer annual; from 6 inches (30 cm) to 4 feet (1.3 m) tall

Place of Origin: eastern North America

Vegetative Characteristics: The growth habit of fall panicum can vary from totally prostate (when growing in poor soil) to upright (when growing in rich soil), and its stems display a distinctive zigzag form. The leaf blades are mostly smooth with a conspicuous white midrib; the leaf sheaths are also smooth and typically are reddish purple. The fibrous root system is extremely tenacious.

Flowers and Fruit: Fall panicum produces wind-pollinated flower panicles from late summer through fall, followed by loose, spreading seed heads that turn purple and then brown with the first frost.

Germination and Regeneration: The seeds germinate readily on bare ground.

Habitat Preferences: This species is very common in the urban environment, especially in sunny sites with compacted soil, including small pavement cracks and under highway guardrails where blacktop and concrete come together.

Ecological Function: Disturbance-adapted colonizer of bare ground.

Cultural Significance: The presence of fall panicum in cracks in parking lots and sidewalks creates the impression of neglect.

Related Species: Switchgrass (*Panicum virgatum* L.) is a large, clump-forming perennial grass native to North America. It often grows in dry soils along sandy roadsides and at the upland edge of salt marshes and other wetlands. Its dead leafy stems persist through the winter. Because of its ability to produce abundant bio-mass on marginal land—it can grow up to 7 feet (2.1 m) tall—switchgrass is being promoted for cultivation on marginal land as a source of cellulosic ethanol. Many cultivars of this species have been selected for ornamental purposes.

Fall panicum growth habit

Fall panicum taking over an abandoned
parking lot

Fall panicum
in its urban
niche

Fall panicum
in bloom

Switchgrass growing in a median strip

Switchgrass in bloom

Phalaris arundinacea L.   Reed Canarygrass

SYNONYM: lady grass

LIFE FORM: **herbaceous perennial**; up to 5 feet (1.6 m) tall

PLACE OF ORIGIN: Eurasia and North America; most of the plants growing in eastern North America originated in Europe.

VEGETATIVE CHARACTERISTICS: The stout, upright stems of reed canarygrass are bluish green, and the smooth leaves appear to be arranged mostly on one side of the stem. The blades are broad and flat, 4–10 inches (10–25 cm) long by 0.5–1.5 inches (1–4 cm) wide, and have a pointed tip and rough edges. The leaves are rolled as they emerge from the bud.

FLOWERS AND FRUIT: Tall flower stalks extend well above the leaves from mid-May through August. The actual inflorescence, which is about 6 inches (15 cm) long, is initially compressed into a tight spike but later opens up to form a loose panicle, and, following wind-pollination, matures into a straw-colored seed head.

GERMINATION AND REGENERATION: The seeds germinate readily; established plants spread by means of thick underground rhizomes that produce dense clusters of stems.

HABITAT PREFERENCES: Reed canarygrass prefers moist, sunny sites and can become dominant in degraded wetlands. It is common along the margins of freshwater marshes, streams, and ponds; in moist woods; and in roadside drainage ditches.

ECOLOGICAL FUNCTIONS: Nutrient absorption in wetlands; tolerant of roadway salt and compacted soil; food and habitat for wildlife; erosion control on stream banks.

CULTURAL SIGNIFICANCE: Although many states list reed canarygrass as an invasive species, it is still widely cultivated in the Northeast as a forage crop for cattle. The old-fashioned variegated variety known as "ribbon grass" (the cultivar 'Picta') is widely grown as an ornamental.

Reed canarygrass in bloom

Reed canarygrass
rhizomes

Reed canarygrass foliage

Reed canarygrass growth habit

Reed canarygrass (*foreground*)
in a degraded wetland with
yellow flag iris (*midground*) and
common reed (*background*)

Phleum pratense L.   Timothy

SYNONYMS: herd's grass, cat's tail

LIFE FORM: herbaceous perennial; 3–4 feet (1–1.3 m) tall

PLACE OF ORIGIN: Europe

VEGETATIVE CHARACTERISTICS: The stems of timothy have a distinct, bulblike thickening at the base from whence its fibrous root system emerges. The smooth, gray-green leaves have flat blades—3–9 inches (8–23 cm) long—that taper to a sharp point and are widely separated from one another on the stem. The new leaves are rolled in the bud (as opposed to being folded) as they emerge.

FLOWERS AND FRUIT: In June and July timothy produces a compact, cylindrical spike of flowers 2–4 inches (5–10 cm) long and less than 0.5 inch (1 cm) wide at the end of a tall, leafless stalk. The inflorescences are purplish and remarkably soft to the touch; following wind-pollination, they mature into light brown seed heads. Individual shoots die after they set seed.

GERMINATION AND REGENERATION: The seeds germinate readily in the fall, shortly after dispersal; established plants increase in size through the production of new shoots from the swollen base; the entire clump typically survives 4 to 5 years.

HABITAT PREFERENCES: Timothy grows best in moist, nutrient-rich soil. In the urban landscape it typically has a scattered distribution in minimally maintained grasslands and meadows; along roadsides; and on the margins of freshwater wetlands, ponds, and streams. It is often a component of erosion control seed mixes.

ECOLOGICAL FUNCTIONS: Food and habitat for wildlife; erosion control on slopes.

CULTURAL SIGNIFICANCE: Timothy is important as forage and as a pasture grass for making hay. Between 1870 and 1910, before the internal combustion engine was invented, it was a major cash crop because it fueled the horses that propelled machines and carriages. The first U.S. record of timothy dates to 1711 when Jonathan Herd found it growing along the Piscataqua River in New Hampshire, most likely as an accidental introduction from England. Its use for hay was first publicly promoted in 1720 by a farmer named Timothy Hanson; hence the common name. The naturalist John Bartram was another early proponent of planting timothy for feeding livestock.

SIMILAR SPECIES: The foxtail grasses (*Setaria* spp.) have bristles associated with their spikelets that give their seed heads a coarse, fuzzy appearance quite unlike the soft, smooth spikes of timothy.

Timothy
inflorescences

Bulb at base
of timothy
stalk

Bluish green
timothy foliage

Mature timothy seed head

Timothy in bloom

Phragmites australis (Cav.) Trin. ex Steud.　Common Reed

SYNONYMS: *Phragmites communis, P. maximus, P. phragmites,* reedgrass, giant reed, marsh reed

LIFE FORM: herbaceous perennial; up to 15 feet (5 m) tall

PLACE OF ORIGIN: Asia, Europe, and North America

VEGETATIVE CHARACTERISTICS: This tall, long-lived grass produces round, hollow stems and alternate, "spikey" leaves that are conspicuously arranged in a single plane. The smooth, blue-green leaf blades are 1–2 feet (30–60 cm) long by about an inch (2.5 cm) wide and have rough margins and prominent veins.

FLOWERS AND FRUIT: The inflorescence is a conspicuous plumelike panicle, 6–15 inches (15–40 cm) long, that is purple at first and, following wind-pollination, changes to tawny brown as the seeds mature. The dry stalks with their tan leaves and fluffy seed heads remain standing through winter and can become a fire hazard during periods of drought.

GERMINATION AND REGENERATION: The seeds are dispersed by wind and can germinate under a variety of moisture and light conditions. More commonly common reed spreads by rhizomes and can form huge populations of genetically uniform stems.

HABITAT PREFERENCES: Common reed is a disturbance-adapted plant that can dominate sunny wetlands—both fresh and salt water—throughout the temperate world. It is common in moist soils wherever drainage has been impeded by roadway construction; in vacant lots and rubble dumps; at the margins of freshwater wetlands, ponds, and streams; and in brackish tidal marshes.

ECOLOGICAL FUNCTIONS: Nutrient absorption (mainly nitrogen and phosphorus) in degraded wetlands; tolerant of roadway salt; food and habitat for wildlife (muskrats); stream, river and shoreline stabilization.

CULTURAL SIGNIFICANCE: Europeans and Middle Easterners used the dry stalks of common reed as building material (mainly for roof thatching) and fed the young shoots to domestic animals. In Russia the plant has been used as a source of cellulose for making paper. Native Americans used many parts of the plant for food: the boiled rhizomes in winter, the young shoots as a vegetable in spring, and the seeds as a grain in fall. In the Southwest the stems were used in the construction of adobe huts and as shafts for arrows. Europeans used the stems to make quills for pens. Common reed is widely planted throughout the temperate world today for soil stabilization and, in constructed wetlands, for the tertiary treatment of wastewater from sewage treatment plants (phytoremediation).

RELATED SPECIES: *Phragmites australis* subspecies *americanus* is native to the East Coast but is relatively rare and not nearly as salt-tolerant as the aggressively spreading variety (**P. a.** subspecies *australis*) that seems to have been unintentionally introduced from Europe in the early 1800s. Indeed, the European variety of *Phragmites* is one of the most widely distributed plants in the world and is listed as an invasive species by many states.

Common reed in
the New Jersey
Meadowlands

Common reed is remarkably drought tolerant

Common reed growing in a stagnant
urban wetland

Phragmites australis variety *gigantissima*
growing up to 20 feet (6 m) tall in Boston's
Back Bay Fens

Common reed flowering shoots

Poa annua L.   Annual Bluegrass

Synonyms: spear grass, dwarf meadow grass, winter grass, six-weeks grass, causeway grass, walk-grass

Life Form: annual; up to 1 foot (0.3 m) tall

Place of Origin: Europe

Vegetative Characteristics: This small, tufted grass has smooth stems that root wherever they touch the ground. The leaf blades are up to 2 inches (5 cm) long, smooth, and light yellowish green with flattened sheaths; the new leaves are folded in bud. The whole plant dies in the midsummer heat. Members of the genus *Poa* can be distinguished from other grasses by their prow-shaped leaf tips.

Flowers and Fruit: In early spring and again in the fall, annual bluegrass produces short, dense panicles of silvery white, wind-pollinated flowers 1–3 inches (2.5–7 cm) long. It is an opportunist species that can mature its seeds within a few days of pollination (hence the name "six-weeks grass").

Germination and Regeneration: The seeds germinate in late summer, early fall, or spring, depending on the location and the weather. Seeds buried in the soil can survive for up to 30 years.

Habitat Preferences: Annual bluegrass is common in compacted lawns and walkways, minimally maintained public parks, vacant lots, rubble dumps, small pavement openings and cracks, and unmowed highway banks and median strips.

Ecological Functions: Disturbance-adapted colonizer of bare ground; remarkably tolerant of roadway salt and soil compaction.

Cultural Significance: This inconspicuous species is one of the most common grasses in the urban environment.

Related Species: Canada bluegrass (*Poa compressa* L.) is a spring-flowering European perennial with a clump-forming growth habit and fine, narrow, blue-green foliage. It typically grows to between 1 and 2 feet (30–60 cm) tall with wiry stems that are distinctly flattened rather than round. The stems are topped with short, compact flower panicles that are 1–3 inches (2.5–7.5 cm) long. Canada bluegrass is common on dry, infertile sites and typically goes dormant in summer.

Annual bluegrass in its urban niche

Prow-shaped leaf tip typical of the genus *Poa*

Annual bluegrass growing in a cobblestone drain

Canada bluegrass stems and foliage

Canada bluegrass in flower

Poa pratensis L.   Kentucky Bluegrass

Synonyms: smooth meadow grass, Junegrass

Life Form: cool-season, **semi-evergreen perennial**; up to 2 feet (60 cm) tall

Place of Origin: western North America and Europe

Vegetative Characteristics: Kentucky bluegrass produces tufts or clumps of smooth leaves that are folded in the bud and blades about 4 inches (10 cm) long. The tips of the blades are shaped like the prow of a boat, a feature common to all members of the genus *Poa*.

Flowers and Fruit: Kentucky bluegrass produces open panicles of green, wind-pollinated flowers in May or June (hence the name "Junegrass"); the flower heads occupy the top 20% or so of the flower stalk, which typically is about 2 feet (60 cm) tall. The stalks are round or oval in cross section with distinct, blackish "joints" at the upper, leafless nodes. The seed heads turn tan and then brown as they mature. In some varieties of Kentucky bluegrass the flowers develop apomictically.

Germination and Regeneration: The seeds germinate in spring or fall in moist, sunny or shady locations; established plants produce creeping rhizomes that allow the clumps to expand quickly. Plants typically go dormant in summer when exposed to warm temperatures, drought, or both.

Habitat Preferences: Given its reputation for high maintenance—lots of water and good soil—it is surprising to find Kentucky bluegrass growing spontaneously in a variety of disturbed sites, including the edges of woodlands, along compacted walkways, and as a component of urban meadows. It prefers limestone soils and is tolerant of road salt.

Environmental Functions: Food and habitat for wildlife; erosion control on slopes.

Cultural Significance: Kentucky bluegrass has a long history of cultivation in pastures and is the most widely grown lawn grass in the Northeast. Its popularity stems from its deep green color and its capacity to spread quickly and produce thick clumps. Although it is apparently native to parts of western North America, Kentucky bluegrass may have been introduced into eastern North America from Europe around 1685 by William Penn. The musical genre "bluegrass" was named for this species.

Related Species: Wood bluegrass (*Poa nemoralis* L.) is an evergreen European species that grows to about 2 feet (60 cm) tall and is typically found in the understory and along the edges of woodlands. Because it does not produce rhizomes or stolons, wood bluegrass has a more clumped growth habit than Kentucky bluegrass and does not tolerate mowing. Its densely clustered flower stalks are produced in June and arch over gracefully as they mature, producing a beautiful effect in the summer landscape.

Unmowed Kentucky bluegrass

Kentucky bluegrass in flower

Wood bluegrass growth habit

Wood bluegrass coming up
in early spring

Wood bluegrass in winter

Wood bluegrass in flower

Setaria viridis (L.) Beauv.   Green Foxtail

Synonym: green brittle grass

Life Form: summer annual; from 4 inches (10 cm) to 3 feet (1 m) tall

Place of Origin: Eurasia

Vegetative Characteristics: Green foxtail produces vigorous new shoots, or tillers, at the base that form a dense clump. The light green leaves are rolled in the bud, and the blades can be up to 10 inches (25 cm) long.

Flowers and Fruit: The light green, wind-pollinated flowers are produced in mid-to-late summer. They are about 3 inches (8 cm) long and are arranged in tight, upright spikes at the end of tall, leafless stalks. The individual flowers are subtended by numerous stiff bristles. The whole plant dies in the fall, leaving the tawny brown seed heads standing tall above a skirt of withered leaves.

Germination and Regeneration: Seeds are dispersed when the mature heads shatter; they germinate in late spring or early summer when the soil starts to warm up.

Habitat Preferences: Green foxtail is quite common in the urban environment, especially in compacted soil and full sun. It grows in small pavement cracks and openings, minimally maintained public parks and landscape planting beds, and unmowed highway banks and median strips.

Ecological Functions: Tolerant of roadway salt and compacted soil; food and habitat for wildlife.

Cultural Significance: Green foxtail is the wild progenitor of foxtail millet (*Setaria italica*), which the Chinese domesticated some 8,000 years ago. Green foxtail has many uses in Asian and European traditional medicine and was intentionally introduced into the United States as a forage grass in the early 1800s.

Related Species: Yellow foxtail (*Setaria glauca* (L.) Beauv.) produces numerous tillers and forms clumps that are about 3 feet (1 m) tall; its upright seed heads, up to 6 inches (15 cm) long, are distinctly yellow and bristly. It is common along roadways and abandoned grasslands and stands out in August when its flower heads create a yellow haze on the landscape. **Giant foxtail (*Setaria farberi* Herrm.)** was introduced inadvertently in the 1930s from Asia. It is a larger, more robust plant than the other two foxtails, reaching up to 5 feet (1.7 m) tall, and it produces a curving rather than an erect seed head that which can be up to 6 inches (15 cm) long. Both yellow and giant foxtails grow best in moist, rich soil and are more common in the countryside than in the city.

Green foxtail in seed

Green foxtail at the end of its life

Giant foxtail growth habit

Giant foxtail at the end of its life

Giant foxtail in seed

Yellow foxtail in seed

Typha latifolia L.   Common Cattail

SYNONYMS: cat-o-nine-tails, cattail flag, blackamoor, candlewick, water torch, bulrush, punks, corndog grass

LIFE FORM: herbaceous perennial; 6–9 feet (2–3 m) tall

PLACE OF ORIGIN: North America

VEGETATIVE CHARACTERISTICS: Cattail is an erect, grasslike perennial with flat, swordlike leaves arising from the base to create a fanlike appearance. The bluish or gray-green blades and can be up to 3 feet (1 m) long by an inch or so (2.5 cm) wide. New shoots and roots arise in early spring from a stout, creeping rhizome.

FLOWERS AND FRUIT: Common cattail produces brownish flowers in dense, cylindrical spikes from May through July. There are two distinct types of flowers on the stalk: the males—located above—which shed their yellow pollen to the wind and quickly disappear; and the females—located below—which develop into the hotdog-shaped "cat tails" that persist and release their fluffy, wind-dispersed seeds over the course of the fall and winter. A single tightly packed cattail may hold approximately 200,000 seeds!

GERMINATION AND REGENERATION: The seeds germinate in moist soil; established plants are strongly rhizomatous and form large stands.

HABITAT PREFERENCES: Cattails grow best along the sunny margins of brackish marshes, freshwater wetlands, and ponds. They are also common in drainage ditches along the margins of highways.

ECOLOGICAL FUNCTIONS: Nutrient absorption in wetlands; tolerant of roadway salt; food and habitat for wildlife; stream and river bank stabilization.

CULTURAL SIGNIFICANCE: The peeled rhizomes and tender young shoots can be eaten raw in salads or cooked as a vegetable. Native Americans used all parts of the plant to treat a variety of skin injuries and ailments as well as diarrhea and dysentery. The plant also has a number of industrial uses based on its processed fibers.

A cattail marsh
in Boston

Cattail foliage

Cattail and purple loosestrife

Cattail flowers in
spring, with males
located above female

Cattails
dispersing
their seeds

APPENDIX 1. Plants Treated in This Book That Are Included in Dioscorides' *De Materia Medica*

The Greek physician Pedanius Dioscorides (AD 40–90) practiced medicine in Rome under the emperor Nero and wrote the five-volume *De Materia Medica*, a compendium of medicinal plants that remained in active use into the 1600s. Modern research has demonstrated that many of the plants Dioscorides mentioned contain medically active compounds and were effective herbal remedies in their day. Among the plants mentioned in the earliest known edition of this book, from the fifth century (Gunther, 1959), are the following:

Achillea millefolium	Yarrow
Arctium minus	Burdock
Calystegia sepium	Hedge Bindweed
Capsella bursa-pastoris	Shepherd's Purse
Chelidonium majus	Greater Celandine
Cynanchum rossicum	Pale Swallowwort
Daucus carota	Wild Carrot
Glechoma hederacea	Ground Ivy
Hypericum perforatum	Common St. Johnswort
Iris pseudacorus	Yellow Flag Iris
Leonurus cardiaca	Motherwort
Medicago sativa	Alfalfa
Polygonum persicaria	Ladysthumb
Populus alba	White Poplar
Portulaca oleracea	Purslane
Rhamnus cathartica	Common Buckthorn
Saponaria officinalis	Bouncing Bet
Solanum nigrum	Black Nightshade
Sonchus oleraceus	Annual Sowthistle
Tussilago farfara	Coltsfoot
Urtica dioica	Stinging Nettle
Verbascum thapsus	Common Mullein

APPENDIX 2. European Plants Listed by Josselyn as Growing Spontaneously in New England in the Seventeenth Century

Plants listed under the category "Of such plants as have sprung up since the English planted and kept cattle in New England":

Capsella bursa-pastoris	Shepherd's Purse
Malva neglecta	Common Mallow
Plantago major	Broadleaf Plantain
Polygonium aviculare	Prostrate Knotweed
Rumex crispus	Curly Dock
Senecio vulgaris	Groundsel
Solanum nigrum	Black Nightshade
Sonchus oleraceus	Annual Sowthistle
Stellaria media	Chickweed
Taraxicum officinale	Dandelion
Urtica dioica	Stinging Nettle

Josselyn mistakenly listed the following plants as native to both England and New England

Arctium minus	Common Burdock
Chelidonium majus	Greater Celandine
Chenopodium album	Lambsquarters
Glechoma hederacea	Ground Ivy
Hypericum perforatum	St. Johnswort
Linaria vulgaris	Yellow Toadflax
Polygonum persicaria	Ladysthumb
Portulaca oleracea	Purslane
Tanacetum vulgare	Common Tansy
Veronica arvensis	Corn Speedwell

Source: From *New-England's Rarities* (1672) by John Josselyn with modern identifications provided by Tuckerman (1865).

APPENDIX 3. Shade-Tolerance Ratings
of the 32 Trees Covered in This Book

Shade-intolerant species (59.4%)

Acer negundo	Box Elder
*Acer pseudoplatanus**	Sycamore Maple
Acer saccharinum	Silver Maple
Ailanthus altissima	Tree-of-Heaven
*Albizia julibrissin**	Silktree
Alnus glutinosa	Black Alder
Betula nigra	River Birch
Betula populifolia	Gray Birch
Catalpa speciosa	Northern Catalpa
Gleditsia triacanthos	Honey Locust
Morus alba	White Mulberry
*Paulownia tomentosa**	Princess Tree
Populus alba	White Poplar
Populus deltoides	Eastern Cottonwood
Populus tremuloides	Quaking Aspen
Quercus palustris	Pin Oak
Rhus glabra	Smooth Sumac
Rhus typhina	Staghorn Sumac
Robinia pseudoacacia	Black Locust

Moderately shade-tolerant species (25%)

Acer rubrum	Red Maple
Celtis occidentalis	Hackberry
Malus pumila	Common Apple
Prunus serotina	Black Cherry
Prunus virginiana	Choke Cherry
Quercus rubra	Red Oak
Ulmus americana	American Elm
Ulmus pumila	Siberian Elm

Very shade-tolerant species (15.6%)

Acer platanoides	Norway Maple
Frangula alnus	Glossy Buckthorn
*Phellodendron amurense**	Amur Corktree
Rhamnus cathartica	Common Buckthorn
*Taxus cuspidata**	Japanese Yew

Source: Based on data listed in *Michigan Trees* by B. V. Barnes and W. H. Wagner, University of Michigan Press, 2004.

* species not listed in *Michigan Trees*

APPENDIX 4. Species Suitable for a Cosmopolitan Urban Meadow

The three plant families that make up this list—the grasses, the legumes, and the composites—often grow together in typical meadow conditions. The grasses have fibrous roots that bind the soil and quickly generate organic matter; the legumes have ropy roots with nitrogen-fixing bacteria that enrich the soil; the composites have a strong taproot that is well adapted to penetrating compacted soil. Growing together, they represent a long-term solution to problematic urban soils that typically cannot support ornamental landscapes.

Species	Common Name	Family
Achillea millefolium	Yarrow	Asteraceae
Cichorium intybus	Chicory	Asteraceae
Leucanthemum vulgare	Oxeye Daisy	Asteraceae
Symphyotrichum pilosum	White Heath Aster	Asteraceae
Tanacetum vulgare	Tansy	Asteraceae
Coronilla varia	Crownvetch	Fabaceae
Lotus corniculatus	Birdsfoot Trefoil	Fabaceae
Trifolium hybridum	Alsike Clover	Fabaceae
Trifolium repens	White Clover	Fabaceae
Eragrostis spectabilis	Purple Lovegrass	Poaceae
Festuca rubra	Red Fescue	Poaceae
Lolium arundinacea	Tall Fescue	Poaceae
Poa compressa	Canada Bluegrass	Poaceae

APPENDIX 5. Key Characteristics of Important Plant Families

Aceraceae (recently merged into the Sapindaceae): Maple Family
- Opposite leaves with lobed margins; they may be simple or compound.
- Flowers are wind-pollinated and typically, but not always, unisexual.
- Winged seeds (samaras) are produced in pairs.

Asteraceae (Compositae): Sunflower Family
- Many individual flowers (florets) are combined on a common receptacle to form a composite inflorescence.
- Two basic types of flowers make up the composite heads: ray or ligulate florets with strap-shaped, petal-like appendages, often infertile; tube-shaped disk florets that lack petal-like appendages and are usually fertile (i.e., seed producing).
- Some, such as the oxeye daisy, have both ray and disk florets.
- Some, such as dandelion, chicory, and prickly lettuce, have only ray florets with conspicuous "petals" and have milky sap.
- Some, such as burdock, groundsel, and pineapple weed, have only disk florets without conspicuous "petals" and have clear sap.
- Flowers pollinated by insects or self-pollinated.

Brassicaceae (Cruciferae): Mustard Family
- Herbaceous plants, often producing basal rosettes.
- Simple, alternate leaves are often divided, cleft, or lobed.
- Flowers have 4 petals forming a cross (mainly yellow or white).
- Flowers pollinated by insects or self-pollinated.
- Seedpod is a 2-chambered capsule that is either long and narrow (silique) or more or less rounded (silicle). The seedpods often explode at maturity, leaving only a clear membrane (the placenta) behind. The sap contains pungent oils.

Caryophyllaceae: Pink Family

- Leaves are opposite, entire, and joined together at the base of the stem to form swollen nodes.
- Flowers have 5 petals, often notched at the tip.
- Sepals are often united to form an inflated tube.
- Flowers pollinated by insects or self-pollinated.
- Fruits consist of a dry capsule filled with small seeds.

Euphorbiaceae: Spurge Family

- Flowers often have colored bracts that look like petals.
- Flowers pollinated by wind or by insects.
- Plants always exude milky sap (rich in latex) when broken.
- Leaves and branching are alternate.
- Plants are often, but not always, toxic to humans.

Fabaceae (Leguminosae): Pea or Bean Family

- Consists of three subfamilies: the Mimosadae, with "powder-puff" flowers; the Caesalpinoideae with separate, more or less equal petals; and the Papalinoideae, with the typical pea-type flower consisting of 5 petals: one is upright (the banner), 2 are lateral on the sides (the wings), and 2 are fused and project outward (the keel).
- Flowers pollinated by bees and other insects.
- All subfamilies produce a fruit that is a pealike pod (a legume) containing one or more seeds and that splits open on two sides.
- Alternate leaves are often pinnately compound.
- Most members of the subfamily Papalinoideae have root nodules that fix nitrogen.

Lamiaceae (Labiatae): Mint Family

- Stems are square and foliage is aromatic.
- Opposite, simple leaves have toothed margins.
- Flowers have 5 fused petals that form 2 prominent lips: an upper one with 2 lobes and a lower one with 3 lobes.
- Flowers typically pollinated by insects.

Poaceae (Graminae): Grass Family

- Stems are round with distinct nodes and hollow internodes.
- Leaves are alternate and often arranged in a single plane.
- Leaves consist of a blade that projects out from the stem and a sheath that wraps around the stem; the blades have parallel venation.
- Flowers are subtended by bracts and aggregated into clusters known as spikelets.
- Flower spikelets lack distinctive petals and sepals and are wind-pollinated.
- New shoots arise from rhizomes or stolons.

Polygonaceae: Smartweed Family

- Stem is swollen at the leaf nodes, forming distinct "knots."
- Leaves are alternate, simple, without teeth along their edges.
- Membranaceous stipules (ochrea) encircle the stem at the nodes.
- Flowers are small with reddish sepals and without petals.
- Flowers pollinated by wind or by insects.
- One-seeded fruits (achenes) usually have 3 distinct wings to facilitate wind or water dispersal.

Rosaceae: Rose Family

- Flowers with 5 petals, 5 sepals, and numerous stamens.
- Flowers typically insect-pollinated.
- Leaves are alternate, can be simple or compound.
- Leaves and/or leaflets are often oval with serrated margins and leaflike stipules at their base.

Scrophulariaceae: Figwort Family

- Flowers typically have 5 petals, which are usually united to form a corolla tube with 2 prominent lips: an upper one with 2 lobes and a lower one with 3 lobes.
- Stamens are attached to the inside of the corolla.
- Flowers typically insect-pollinated.
- Fruit is typically a dry capsule filled with small seeds.

Vitaceae: Grape Family

- Climbing vines with tendrils that coil or produce adhesive disks at their tips.
- Leaves are alternate, deeply lobed or compound.
- Tendrils and flower clusters are produced opposite the leaves.
- Flowers typically insect-pollinated.
- Clusters of fleshy berries contain relatively few seeds.

GLOSSARY*

Achene A single-seeded, dry fruit that remains closed at maturity (e.g., the "seeds" of buttercups and cinquefoils).

Actinobacteria A type of filamentous, gram-positive bacteria. The genus *Frankia* forms nitrogen-fixing root nodules in symbiosis with a variety of dicotyledons (e.g., alder, autumn olive). *Contrast* Rhizobium

Adventitious bud or shoot A bud or shoot that arises from any region of the plant other than the leaf axil, as from a root or a leaf.

Adventitious root A root that originates from stem or leaf tissue rather than from another root.

Adventive Refers to an introduced or nonnative species with only a limited or temporary distribution in a given area; not widespread. *Contrast* Invasive; Naturalized

Allelopathy Release of a chemical compound by one plant that inhibits the growth of another plant.

Alternate Refers to leaves or other organs arranged singly at a node. *Contrast* Opposite; Whorled

Angiosperm A plant that produces flowers with ovules enclosed in an ovary. *Contrast* Gymnosperm

Annual A plant that completes its life cycle in 1 year, including germination, flowering, seed set, and death. *See* Summer annual; Winter annual

Anther The enlarged, terminal portion of the stamen that produces pollen.

Apical Located at the tip of an organ such as a leaf, root, or shoot. *Contrast* Basal; Lateral

---

* Definitions are based on a number of sources, the most significant are H. A. Gleason and A. Cronquist, *Manual of Vascular Plants of Northeastern United States and Adjacent Canada*, 2nd ed. (Bronx, N.Y.: New York Botanical Garden, 1991); R. H. Uva, J. C. Neal, and J. M. DiTomaso, *Weeds of the Northeast* (Ithaca, N.Y.: Cornell University Press, 1997); M. Hickey and C. King, *The Cambridge Illustrated Glossary of Botanical Terms* (Cambridge: Cambridge University Press, 2000); and J. M. DiTomaso and E. A. Healy, *Weeds of California and Other Western States* (Berkeley: University of California Press, 2007).

Apomixis The production of seeds without fertilization; a form of asexual reproduction.

Archaeophyte A plant introduced into European agriculture prior to AD 1500. *Contrast* Neophyte

Aril A fleshy, often brightly colored covering on some seeds (e.g., those of the yew tree).

Asexual reproduction *See* Vegetative reproduction

Auricle In grasses, a small, projecting lobe or earlike appendage located where the blade meets the sheath; in broadleaf plants, an earlike lobe that protrudes from the base of the leaf blade.

Awn The slender bristle on a grass floret.

Axil The angle formed by the upper side of a leaf and the stem. The position on a stem above the point of attachment of a leaf where a bud is located is the "leaf axil." *See* Subtend

Basal Located at the base of an organ such as a stem, leaf, or flower. *Contrast* Apical; Lateral

Berry A fleshy, indehiscent fruit with 1 or more seeds embedded in pulp.

Biennial A plant that requires 2 years to complete its life cycle. During the first season the seed germinates and produces a rosette of leaves; the following year it flowers, sets seed, and dies. *Contrast* Annual

Bipinnate Refers to leaves that are branched twice, with leaflets on the second-order branches. *See* Compound leaf; *Contrast* Pinnate

Bisexual A flower having both male and female parts. *See* Perfect flower; *Contrast* Unisexual

Blade The expanded, flattened portion of a leaf; located above the petiole in broadleaf plants and above the sheath in grasses.

Bloom A waxy powder that covers the surface of a leaf or fruit, making it appear bluish or whitish. *See* Glaucous

Bolt To produce erect, flowering stems from a basal rosette of leaves; typically associated with annual or biennial species.

Bract A reduced, leaflike structure located below a flower or inflorescence.

Bud A young shoot protected by scale leaves from which flowers or leaves develop; typically located in the axil of a leaf or bract.

Bulb A short underground shoot with leaves modified to act as food storage organs.

Bur A fruiting structure covered with spines or prickles, typically dispersed as a unit.

Calyx Collectively, the sepals of a flower. *Contrast* Corolla

Capsule A dry fruit that splits open at maturity to release its seeds.

Carbohydrate An organic compound composed of carbon, hydrogen, and oxygen, such as a sugar or starch; one of the products of photosynthesis.

Carpel The female organ at the center of a flower, consisting of an ovary, a style, and a stigma. *See* Pistil

Catkin A pendulous, spikelike inflorescence of unisexual flowers that lack petals; found in the Betulaceae.

Circumboreal Occurring all the way around the North Pole, encompassing the northern parts of Europe, Asia, and North America.

Clasping Refers to the base of a leaf blade that surrounds the stem to which it is attached.

Cleistogamous flower A self-pollinating flower that produces seeds without opening, as in violets and jewelweed.

Clone In plants, a group of genetically identical individuals produced by means of vegetative reproduction from a single parent.

Collar The outer side of a grass leaf located at the junction of the blade and sheath; also the base of a tree or shrub where new shoots originate. *See* Crown

Colonizer *See* Pioneer

Composite flower head The dense inflorescence of the Asteraceae composed of individual florets carried on a receptacle subtended by bracts.

Compound leaf A leaf composed of 2 or more leaflets. Once-compound leaves have leaflets arranged along an unbranched petiole or rachis; twice-compound leaves have leaflets arranged along a branched petiole or rachis. *See* Bipinnate; Palmate; Pinnate

Cone The organ of sexual reproduction in gymnosperms; woody female cones produce seeds and transitory male cones produce pollen. *Contrast* Flower

Conical Cone shaped.

Conifer A seed-bearing plant that produces cones; part of the gymnosperm group.

Cordate Heart shaped.

Corolla The collective term for the petals of a flower, either separate or fused. *Contrast* Calyx

Corymb An inflorescence with branches arising at different points on a stem but reaching more or less the same height at maturity, producing a flat-topped appearance. *Contrast* Umbel

Cotyledon The first leaf (monocotyledon) or pair of leaves (dicotyledon) produced by a seedling.

Crown The base of a perennial plant, located at or just below ground level, where new shoots originate. *See* Collar; Stump sprout

Culm A jointed stem, especially the flowering stem of grasses.

Cultivar A individual perennial plant or strain of annual plant that has been selected and propagated for its unique characteristics. The word is a contraction of the phrase "cultivated variety."

Deciduous Dropped at senescence, as leaves in autumn or petals after flowering.

Dehiscent Splitting open at maturity, as a pod releasing its seeds.

Deltoid Shaped more or less like an equilateral triangle.

Dicot Short term for dicotyledon.

Dicotyledon A broadleaf flowering plant (i.e., an angiosperm) that is characterized by seedlings with 2 cotyledons (along with a number of other features). *Contrast* Monocotyledon

Dimorphic Producing two different forms of the same organ, as the leaves of perennial vines.

Dioecious Producing male and female flowers on different plants of the same species. *Contrast* Monoecious

Disk floret The central, tubular flower of some members of the Asteraceae with both male and female organs. *Contrast* Ray floret

Dissected Divided into many slender, irregular segments but not compound; typically used to describe leaves or petals.

Dormant Resting or inactive; used to describe seeds or buds that will sprout only after experiencing a period of chilling.

Drupe A fleshy fruit containing a single seed enclosed in a hardened ovary wall (e.g., a cherry or plum).

Elliptical Football shaped; used to describe a leaf that is widest in the middle and narrowing equally toward the ends.

Endemic A plant with a natural distribution restricted to a particular country or region.

Endophyte A bacterium or fungus that lives within a plant for at least part of its life cycle without causing a disease.

Entire The unbroken edge of a leaf or petal with a continuous, untoothed margin.

Escape Any cultivated plant that, under its own power, has spread outside the garden or field in which it was originally planted. *Contrast* Naturalized; Spontaneous; Volunteer

Exotic Nonnative; originating in a foreign country or region. *Contrast* Native

Fastigiate Refers to the growth habit of a tree in which the branches are all more or less erect or ascending, as a Lombardy poplar. *Contrast* Weeping

Fertilization In plants, the fusion of an ovule with a pollen cell to produce a viable seed.

Filament The stalk of a stamen.

Floret A little flower; an individual flower in a flower cluster, as in a grass spikelet or the composite flower head in the Asteraceae.

Flower The organ of sexual reproduction in angiosperms; typically composed of sepals, petals, stamens, and carpels. *Contrast* Cone

Flower head *See* Composite flower head

Frond The leaf of a fern.

Fruit A mature ovary of a plant including the enclosed seeds.

Gametophyte The sexual stage in the life cycle of a plant in which the cells have half the usual number of chromosomes.

Genotype The genetic makeup of an individual organism. *Contrast* Phenotype

Germination The sprouting of a seed to produce a seedling; the production of a pollen tube by a pollen grain.

Glabrous Smooth, without hairs. *Contrast* Pubescent

Glaucous Covered with a waxy, bluish green coating; used to describe leaves or fruits.

Gymnosperm A group of seed plants whose ovules are "naked," i.e., not enclosed in an ovary. *Contrast* Angiosperm

Habitat The environmental conditions in which a plant grows.

Herb A plant with stems that die to the ground at the end of the growing season.

Herbaceous Refers to a plant composed entirely of soft, nonwoody tissue that dies to the ground each year.

Hybrid A plant produced by the sexual crossing of parents belonging to two genetically distinct taxonomic groups.

Hypocotyl On a seedling, the region of the stem between the cotyledons and the primary root.

Indehiscent Refers to a fruit that remains closed at maturity.

Indigenous *See* Native

Inflorescence The grouping or arrangement of flowers on a stem; a cluster of flowers.

Internode The section of a stem between two adjacent nodes.

Invasive Refers to a nonnative species with the capacity to proliferate and spread aggressively into natural habitats or minimally managed landscapes, often displacing native species and reducing biodiversity. *Contrast* Adventive; Naturalized

Lanceolate More or less lance shaped; that is, much longer than wide with a rounded base and a pointed tip.

Lateral Located on the side of an organ; typically used to describe the position of branches, roots, or flowers. *Contrast* Apical; Basal

Leaf A lateral outgrowth from the stem, typically consisting of a petiole and a blade.

Leaflet The leaflike subunit of a compound leaf lacking an associated bud.

Legume A simple, dry fruit that opens lengthwise along two seams and is characteristic of the family Fabaceae, e.g., a string bean or a pea pod.

Lenticle A small pore or opening in the bark that allows gases to pass in and out.

Ligulate floret Refers to a flower type in the Asteraceae with a long, petal-like corolla. *See* Ray floret

Ligule The thin, dry membrane that projects from the top of the leaf sheath in grasses and sedges.

Lobe A rounded segment of a leaf or flower part that is larger than a tooth, typically with the adjoining sinuses extending less than halfway to the midrib; leaf or petal margins are often described as "lobed."

Margin The outer edge of a leaf or petal. *See* Entire; Lobe; Serrate

Membranous Thin, flexible, and transparent.

Monocot Short term for monocotyledon.

Monocotyledon A grasslike flowering plant (i.e., angiosperm) that is characterized by seedlings with 1 cotyledon (along with a number of other features). *Contrast* Dicotyledon

Monoecious Producing separate male and female flowers on the same plant. *Contrast* Dioecious

Mycorrhizae A mutually beneficial, symbiotic association between various fungi and the roots of plants that involves the exchange of minerals and carbohydrates.

Native Occurring naturally in a given region; not introduced into an area as a result of human activity; indigenous. *Contrast* Exotic

Naturalized Refers to an introduced or nonnative species that reproduces on its own and is well established in a region. *Contrast* Adventive; Invasive; Volunteer

Nectary An organ, usually located on a flower but sometimes found on leaves, that produces nectar to attract animals.

Neophyte A plant introduced into European agriculture—mainly from Asia and the Americas—after AD 1500. *Contrast* Archaeophyte

Nitrogen fixation A biochemical process carried out by bacteria in which nitrogen gas in the air is converted into ammonium that is then used to produce amino acids and proteins. *See* Root nodule; Symbiosis

Node The place on a stem where a leaf and its associated bud are attached. *See* Internode

Oblong Much longer than wide, with parallel sides and a more or less rectangular shape.

Ocrea A papery sheath that encloses the stem at the nodes formed by the fusion of 2 stipules; found in members of the Polygonaceae.

Opposite Refers to leaves or other organs arranged directly across from each other at the same node. *Contrast* Alternate; Whorled

Ovary The part of the female flower in angiosperms that contains ovules and develops into a fruit.

Ovate Shaped like a chicken egg, with the larger end closer to the base than the tip.

Ovule A structure within the ovary that, after fertilization, develops into a seed.

Palmate Divided to the base into separate leaflets, all of which arise from the same point at the end of the petiole.

Panicle An inflorescence with a main axis and secondary branches; usually broadest at the base and tapering upward.

Pappus In the family Asteraceae, the tuft of hairs or dry scales on a seed that facilitate wind or animal dispersal.

Pedicel The stalk of a single flower within an inflorescence.

Peduncle The stalk of an inflorescence or a single flower.

Perennial Refers to a plant that lives for more than 2 years. Contrast Annual

Perfect flower A flower with both male and female organs. *See* Bisexual

Petal An individual part of the corolla, usually colored.

Petiole The stalk of a leaf.

Pith The spongy tissue that occupies the central portion of a twig or stem.

pH The scale used to measure the acidity (pH < 7) or alkalinity (pH > 7) of a solution.

Phenotype An observable characteristic or trait of an organism; it can be the product of either genetic or environmental factors or their interaction. *Contrast* Genotype

Photosynthesis The process whereby a green plant converts carbon dioxide and water into sugars and oxygen in the presence of sunlight.

Phytoremediation The use of plants to clean up land contaminated by a variety of human-generated waste products such as heavy metals, petroleum products, acid mine drainage, and excessive amounts of dissolved nitrogen and phosphorus.

Pinnate Refers to a compound leaf with leaflets arranged along both sides of a petiole; they may be odd-pinnate with a single terminal leaflet or even-pinnate with a terminal pair of leaflets. *Contrast* Bipinnate

Pioneer A plant that colonizes bare ground; an early successional species. *Contrast* Invasive; Weed

Pistil The female organ of a flower, consisting of an ovary, style, and stigma. *See* Carpel

Pistillate Refers to flowers having only female organs.

Pith The central, spongy tissue in a stem or root that is surrounded by a cylinder of woody tissue.

Pollen The powdery substance shed from anthers of a flower that contains the male reproductive cells.

Pollination The transfer of pollen from an anther to a stigma.

Pre-adaptation An anatomical or physiological trait that evolved under one set of ecological conditions and, by chance, proves advantageous under a completely different set of circumstances.

Prostrate Lying flat on the ground; used to describe the growth habit of low-growing plants.

Pubescent Covered with short hairs; typically used to describe leaves and stems. *Contrast* Glabrous

Raceme An elongated inflorescence with stalked flowers on an unbranched central axis; flowers develop from the bottom up. *Contrast* Spike

Rachis The central axis of a pinnately compound leaf.

Ray floret One of the outer, irregular flowers in the flower heads of some plants in the Asteraceae that produce a single, straplike "petal." *Contrast* Disk floret

Receptacle The basal part of a flower to which the other flower parts are attached; it is sometimes enlarged and fleshy, as in a strawberry or blackberry.

Recurved Bent or curved downward or backward.

Rhizobium Rod-shaped bacteria that are capable of nitrogen fixation in symbiosis with plants in the family Fabaceae. *Contrast* Actinobacteria

Rhizome A creeping, underground stem that produces new shoots and adventitious roots. *Contrast* Stolon

Root The lower portion of the plant's axis that anchors it in the soil and absorbs nutrients and water. The primary root develops from the embryo, and secondary roots are branches off the primary root.

Root nodule An outgrowth on the roots of certain plants that contains *Rhizobium* or *Frankia* bacteria and is the sites of symbiotic nitrogen fixation.

Root sucker A shoot that arises from an adventitious bud on a root.

Rosette In herbaceous plants, a circular cluster of leaves located at ground level. The stem is compressed and the leaves are separated by very short internodes.

Ruderal Refers to a plant that grows in waste places or requires soil disturbance to become established. From Latin *rudera*, "ruins" or "rubbish," plural of *rudus*, "broken stone." In botanical parlance, a disturbance-adapted species.

Runner *See* Stolon

Samara A dry, indehiscent winged fruit, as maple or ash seeds.

Scale A reduced or rudimentary leaf; usually surrounding a dormant bud.

Scion In grafting, a young shoot with special characteristics that is spliced or grafted onto a rooted understock.

Seed A ripened ovule consisting of a protective coat enclosing an embryo and food reserves; the product of sexual reproduction.

Seed bank The totality of viable seeds that lie buried in the soil at any point in time; they typically germinate following some disturbance of the soil surface.

Seed head An inflorescence bearing mature fruit; especially in the Asteraceae.

Seed leaf *See* Cotyledon

Sepal In a flower, the outermost whorl of leaflike, usually green appendages below the petals. *See* Calyx

Serrate Refers to the margin of a leaf with sharp, forward-pointing teeth. *Contrast* Entire; Lobe

Sessile Attached directly by the base, as a leaf without a petiole or a flower without a pedicel.

Sexual reproduction The production of a plant from seed. *Contrast* Vegetative reproduction

Sheath In grasses, the lower part of a leaf that encloses the stem and the emerging new leaves.

Shoot A young green stem with leaves, flowers, or both.

Simple leaf A leaf blade that consists of a single piece but may be deeply lobed or divided. *Contrast* Compound leaf

Sinuses The indentations between the teeth or lobes on the margin of a leaf.

Species A population of individuals sharing common morphological features and evolutionary history; the basic unit of classification in the nomenclatural hierarchy.

Spike An elongated inflorescence with sessile flowers on an unbranched central axis. *Contrast* Raceme

Spikelet In grasses and sedges, an inflorescence consisting of 1 to many flowers subtended by minute bracts.

Spontaneous Refers to a plant that grows in an area without being cultivated by humans; it may be either a native or introduced species. *Contrast* Escape; Naturalized

Spore The tiny, dustlike reproductive unit of ferns and horsetails.

Sporophyte The mature stage in the life cycle of ferns and horsetails that produces spores.

Stamen The male organ of a flower, consisting of a stalk (filament) and a pollen-producing organ (anther).

Staminate Flowers that have only male organs.

Stigma The apex of the style, usually enlarged and sticky, on which the pollen grains land and germinate. *See* Carpel; Pistil

Stipule A small, leafy outgrowth at the base of a petiole; typically found in pairs.

Stolon A stem that grows horizontally along the surface of the ground, producing roots at its nodes and new plants from its buds. *Contrast* Rhizome

Strobilus The conelike reproductive structure of horsetails.

Stump sprout A shoot that emerges from the base of a tree or shrub, usually after some form of traumatic injury such as logging; a basal sprout. *See* Crown

Style The elongated portion of a carpel, above the ovary, with the stigma at its tip.

Subtend To be located beneath a given structure, as a leaf at the base of a bud. *See* Axil

Succession Changes in the composition of a plant association over time, typically initiated by some form of disturbance. **Early succession** refers to the beginning stages of the process, immediately following a disturbance event; **late succession**

refers to the latter stages of the process when the composition of the vegetation has stabilized.

Summer annual A plant that germinates in late spring or summer and completes its life cycle before winter. *Contrast* Winter annual

Symbiosis A long-lasting association between 2 species of organisms that may or may not be mutually beneficial; literally the word means "living together."

Taproot A thick, downward-growing root with few side branches; adapted to penetrating heavy soils and storing carbohydrates.

Tendril A twining, threadlike structure produced by the stem or leaf of a climbing plant that enables it to cling to an object.

Terminal Located at the tip or apex (e.g., a terminal inflorescence). *Contrast* Lateral

Tiller A lateral shoot that emerges from the base of a grass or other monocot.

Trifoliate A compound leaf consisting of 3 leaflets, as in clover and poison ivy.

Tuber The swollen tip of an underground stem or rhizome; an underground food storage organ that sprouts when growing conditions are favorable (e.g., a potato).

Tubular floret *See* Disk floret

Umbel A cluster of flowers—typically flat topped—with numerous pedicels arising from a common point.

Understock In grafting, the rooted plant (usually a seedling) to which the scion is spliced or grafted.

Undulate Having a wavy, up-and-down margin; used to describe leaves or petals.

Unisexual A flower having organs of only 1 sex. *Contrast* Bisexual

Variegated Having 2 or more colors.

Vegetative reproduction Asexual reproduction by any means other than from seed, including bulbs, rhizomes, stump sprouts, root suckers, and leaves. *See* Clone; *Contrast* Sexual reproduction

Vein An externally visible strand of vascular tissue in a leaf or other flat organ.

Volunteer *See* Spontaneous

Weed A plant that grows rapidly and abundantly in a place where it is not wanted; a plant that aggressively colonizes disturbed habitats. *Contrast* Invasive

Weeping With branches bending over or hanging down, as a weeping willow. *Contrast* Fastigiate

Whorled Having a ring of 3 or more leaves or other organs radiating from a node. *Contrast* Alternate; Opposite

Wing A thin, flat extension or projection from the side of a stem or leaf.

Winter annual A plant that germinates in fall or winter and completes its life cycle the following spring. Winter annuals are tolerant of cold weather, overwinter as rosettes, and grow best in cool, moist conditions. *Contrast* Summer annual

BIBLIOGRAPHY

Anderson, E. 1952. *Plants, Man and Life*. Boston, Mass.: Little, Brown.

Anderson, E., and R. E. Woodson. 1935. The species of *Tradescantia* indigenous to the United States. *Contributions from the Arnold Arboretum of Harvard University*, IX.

Arnold, C. L., and C. J. Gibbons. 1996. Impervious surface coverage: the emergence of a key environmental indicator. *Journal of the American Planning Association* 62(2):243–258.

Baker, H. G. 1974. The evolution of weeds. *Annual Review of Ecology and Systematics* 5:1–24.

Barnes, B. V., and W. H. Wagner Jr. 2004. *Michigan Trees*. Ann Arbor: University of Michigan Press.

Bartram, J. 1992. *The Correspondence of John Bartram, 1734–1777*, ed. E. Berkeley and D. Smith Berkeley. Gainesville: University Press of Florida.

Bormann, F. H., D. Balmori, and G. T. Geballe. 1993. *Redesigning the American Lawn: A Search for Environmental Harmony*. New Haven, Conn.: Yale University Press.

Brown, L. 1977. *Weeds in Winter*. Boston: Houghton Mifflin.

Brown, L. 1979. *Grasses: An Identification Guide*. Boston: Houghton Mifflin.

Burns, R. M., and B. H. Honkala. 1990. *Silvics of North America*. Agriculture Handbook 654. Washington, D.C.: U.S. Department of Agriculture, Forest Service.

Byrne, L. B. 2007. Habitat structure: a fundamental concept and framework for urban soil ecology. *Urban Ecosystems* 10:255–274.

Christopher, T. 2008. Can weeds help solve the climate crisis? *New York Times Magazine*, June 29, 2008.

Clemants, S., and C. Gracie. 2006. *Wildflowers in the Field and Forest*. New York: Oxford University Press.

Coates, P. 2006. *Strangers on the Land: American Perceptions of Immigrant and Invasive Species*. Berkeley: University of California Press.

Crockett, L. J. 1977. *Wildly Successful Plants: A Handbook of North American Weeds*. New York: Macmillan.

Darlington, W. 1859. *American Weeds and Useful Plants*. 2nd edition, revised by G. Thurber. New York: A. O. Moore.

Del Tredici, P. 2007. The role of horticulture in a changing world. In *Botanical Progress, Horticultural Innovation, and Cultural Changes*, ed. M. Conan and W. J. Kress, pp. 259–264. Washington, D.C.: Dumbarton Oaks.

Dirr, M. A. 1998. *Manual of Woody Landscape Plants*. 5th ed. Champaign, Ill.: Stipes Publishing.

DiTomaso, J. M., and E. A. Healy. 2007. *Weeds of California and Other Western States*. Berkeley: University of California Press.

Dunnett, N., and J. Hitchmough, eds. 2004. *The Dynamic Landscape: Design, Ecology and Management of Naturalistic Urban Planting*. London: Spon Press.

Dunnett, N., and N. Kingsbury. 2004. *Planting Green Roofs and Living Walls*. Portland, Or.: Timber Press.

Eastman, J. 1992. *The Book of Forest and Thicket*. Mechanicsburg, Penn.: Stackpole Books.

Eastman, J. 1995. *The Book of Swamp and Bog*. Mechanicsburg, Penn.: Stackpole Books.

Eastman, J. 2003. *The Book of Field and Roadside*. Mechanicsburg, Penn.: Stackpole Books.

Fernald, M. L., and A. C. Kinsey. 1943. *Edible Wild Plants of Eastern North America*. Cornwall-on-Hudson, N.Y.: Idlewild Press.

Fogg, J. M. 1945. *Weeds of Lawn and Garden*. Philadelphia: University of Pennsylvania Press.

Foster, S., and J. A. Duke. 1990. *A Field Guide to Medicinal Plants of Eastern and Central North America*. Boston, Mass.: Houghton Mifflin.

Garber, S. D. 1987. *The Urban Naturalist*. New York: John Wiley and Sons.

George, K., L. H. Ziska, J. A. Bunce, and B. Quebedeaux. 2007. Elevated atmospheric CO_2 concentration and temperature across an urban-rural transect. *Atmospheric Environment* 41:7654–7665.

George, K., et al. 2009. Macroclimate associated with urbanization increases the rate of secondary succession from fallow soil. *Oecologia* 159:637–647.

Gleason, H. A., and A. Cronquist. 1991. *Manual of Vascular Plants of Northeastern United States and Adjacent Canada*. 2nd ed. Bronx, N.Y.: New York Botanical Garden.

Godefroid, S., D. Monbaliu, and N. Koedam. 2007. The role of soil and microclimatic variables in the distribution patterns of urban wasteland flora in Brussels, Belgium. *Landscape and Urban Planning* 80:45–55.

Gould, S. J. 1998. An evolutionary perspective on strengths, fallacies, and confusions in the concept of native plants. *Arnoldia* 58(1):2–10.

Gregg, J.W., C.G. Jones, and T.E. Dawson. 2003. Urbanization effects on tree growth in the vicinity of New York City. *Nature* 424:183–187.

Grieve, M., and C. F. Leyel. 1931. *A Modern Herbal*. New York: Harcourt, Brace.

Grime, J. P. 2001. *Plant Strategies, Vegetation Processes, and Ecosystem Properties*. 2nd ed. New York: John Wiley and Sons.

Grimm, N. B., Stanley H. Faeth, Nancy E. Golubiewski, Charles L. Redman, Jianguo Wu, Xuemei Bai, and John M. Briggs. 2008. Global change and the ecology of cities. *Science* 319(5864):756–760.

Gunther, R. T., ed. 1959. *The Greek Herbal of Dioscorides: Illustrated by a Byzantine, A.D. 512. Englished by John Goodyer, A.D. 1655*. New York: Hafner.

Heiser, C. B. 2003. *Weeds in My Garden*. Portland, Ore.: Timber Press.

Hickey, M., and C. King. 2000. *The Cambridge Illustrated Glossary of Botanical Terms*. Cambridge: Cambridge University Press.

Hu, S.-Y. 2005. *Food Plants of China*. Hong Kong: Chinese University Press.

Hubbard, C. E. 1992. *Grasses*. 3rd ed. London: Penguin Books.

Hutchinson, J. 1946. *Common Wildflowers*. West Drayton, England: Penguin Books.

Hutchinson, J. 1948. *More Common Wildflowers*. West Drayton, England: Penguin Books.

Janzen, D. 1998. Gardenification of wildland nature and the human footprint. *Science* 279:1312–1313.

Jones, P. 1994. *Just Weeds: History, Myths and Uses*. Shelburne, Vt.: Chapters Publishing.

Josselyn, J. 1672 [reprint 1865.] *New England Rarities Discovered*, ed. E. Tuckerman. Boston, Mass.: William Veazie.

Kareiva, P., S. Watts, R. McDonald, and T. Boucher. 2007. Domesticated nature: shaping landscapes and ecosystems for human welfare. *Science* 316:1866–1869.

Kastner, J. 1993. My empty lot. *New York Times Magazine*, October 10, 1993, pp. 22–25, 41–44.

Kenfield, W. G. 1966. *The Wild Gardener in the Wild Landscape*. New York: Hafner.

Kowarik, I. 1990. Some responses of flora and vegetation to urbanization in Central Europe. In *Urban Ecology*, ed. H. Sukopp et al., pp. 45–74. The Hague: Academic Publishers.

Kowarik, I., and S. Körner, eds. 2005. *Wild Urban Woodlands*. Berlin: Springer-Verlag.

Kowarik, I., and A. Langer. 2005. Natur-Park Südgelände: linking conservation and recreation in an abandoned railyard in Berlin. In *Wild Urban Woodlands*, ed. I. Kowarik and S. Körner, pp. 287–299. Berlin: Springer-Verlag.

Kühn, N. 2006. Intentions for the unintentional spontaneous vegetation as the basis for innovative planting design in urban areas. *Journal of Landscape Architecture* (autumn 2006):46–53.

Larson, D., U. Matthes, P. E. Kelly, J. Lundholm, and J. Garrath. 2004. *The Urban Cliff Revolution*. Markham, Ontario: Fitzhenry and Whiteside.

Lundholm, J. T., and A. Marlin. 2006. Habitat origins and microhabitat preferences of urban plant species. *Urban Ecosystems* 9:139–159.

Mack, R. N. 2000. Cultivation fosters plant naturalization by reducing environmental stochasticity. *Biological Invasions* 2:111–122.

Mack, R. N. 2003. Plant naturalizations and invasions in the eastern United States: 1634–1860. *Annals of the Missouri Botanical Garden* 90:77–90.

Mack, R. N., and M. Erneberg. 2002. The United States naturalized flora: largely the product of deliberate introductions. *Annals of the Missouri Botanical Garden* 89:176–189.

Marks, P. L. 1983. On the origin of the field plants of the northeastern United States. *American Naturalist* 122(2):210–228.

Maurer, U., T. Peschel, and S. Schmitz. 2000. The flora of selected urban land-use types in Berlin and Potsdam with regard to nature conservation in cities. *Landscape and Urban Planning* 46:209–215.

Mehrhoff, L. J. 2000. Immigration and expansion of the New England flora. *Rhodora* 102:280–298.

Miller, A. B. 1976. *Shaker Herbs: A History and a Compendium*. New York: Clarkson N. Potter.

Mohan, J. E., Lewis H. Ziska, William H. Schlesinger, Richard B. Thomas, Richard C. Sicher, Kate George, and James S. Clark. 2006. Biomass and toxicity responses of poison ivy (*Toxicodendron radicans*) to elevated atmospheric CO_2. *Proceedings of the National Academy of Sciences (USA)* 103(24):9086–9089.

Muenscher, W. C. 1955. *Weeds*. 2nd ed. New York: Macmillan.

Mühlenbach, V. 1979. Contribution to the synanthropic (adventive) flora of the railroads in St. Louis, Missouri, U.S.A. *Annals of the Missouri Botanical Garden* 66(1):1–108.

Muratet, A., Nathalie Machon, Frédéric Jiguet, Jacques Moret, and Emmanuelle Porcher. 2007. The role of urban structures in the distribution of wasteland flora in the greater Paris area, France. *Ecosystems* 10(4):661–671.

Newcomb, L. 1977. *Newcomb's Wildflower Guide*. New York: Little, Brown.

Page, N., and R. E. Weaver. 1974. Wild plants in the city. *Arnoldia* 34(4):137–232.

Pauly, P. J. 2007. *Fruits and Plains: The Horticultural Transformation of America*. Cambridge, Mass.: Harvard University Press.

Pickett, S. T. A., Mary L. Cadenasso, J. Morgan Grove, Peter M. Groffman, Lawrence E. Band, Christopher G. Boone, William R. Burch Jr., C. Susan B. Grimmond, John Hom, Jennifer C. Jenkins, Neely L. Law, Charles H. Nilon, Richard V. Pouyat, Katalin Szlavecz, Paige S. Warren, and Matthew A. Wilson. 2008. Beyond urban legends: an emerging framework of urban ecology, as illustrated by the Baltimore Ecosystem Study. *BioScience* 59(2):139–150.

Porębska, G., and A. Ostrowska. 1999. Heavy metal accumulation in wild plants: implications for phytoremediation. *Polish Journal of Environmental Studies* 8(6):433–442.

Pyšek, P. 1998. Alien and native species in Central European urban floras: a quantitative comparison. *Journal of Biogeography* 25:155–163.

Pyšek, P., Zdena Chocholoušková, Antonín Pyšek, Vojtěch Jarošík, Milan Chytry, and Lubomír Tichy. 2004. Trends in species diversity and composition of urban vegetation over three decades. *Journal of Vegetation Science* 15:781–788.

Rehder, A. 1946. On the history of the introduction of woody plants into North America. *Arnoldia* 6(4–5):13–28.

Riddle, J. M. 1999. *Eve's Herbs: A History of Contraception and Abortion in the West*. Cambridge, Mass.: Harvard University Press.

Royer, F., and R. Dickinson. 1999. *Weeds of the Northern U.S. and Canada*. Edmonton, Alberta: Lone Pine Publishing and University of Alberta Press.

Sagoff, M. 2005. Do non-native species threaten the natural environment? *Journal of Argicultural and Environmental Ethics* 18:215–236.

Salisbury, E. 1961. *Weeds and Aliens*. London: Collins.

Sieghardt, M., Erich Mursch-Radlgruber, Elena Paoletti, Els Couenberg, Alexandros Dimitrakopoulus, Francisco Rego, Athanassios Hatzistathis, and Thomas Barfoed Randrup. 2005. The abiotic urban environment: impact of urban growing conditions on urban vegetation. In *Urban Forests and Trees*, ed. C. C. Konijnendijk, Kjell Nilsson, Thomas Randrup, and Jasper Schipperijn, pp. 281–323. Berlin: Springer.

Smith, B. 1943. *A Tree Grows in Brooklyn*. New York: Harper and Brothers.

Somers, P., R. Kramer, K. Lombard, and B. Brumback. 2006. *A Guide to Invasive Plants in Massachusetts*. Westboro: Massachusetts Division of Fish and Wildlife.

Sorrie, B. A. 2005. Alien vascular plants in Massachusetts. *Rhodora* 107:284–329.

Sorrie, B. A., and P. Somers. 1999. *The Vascular Plants of Massachusetts: A County Checklist*. Westboro: Massachusetts Division of Fisheries and Wildlife.

Stalter, R. 2004. The flora of the High Line, New York City, New York. *Journal of the Torrey Botanical Society* 131(4):387–392.

Sukopp, H. 2004. Human-caused impact on preserved vegetation. *Landscape and Urban Planning* 68:347–355.

Tallamy, D. W. 2007. *Bringing Nature Home*. Portland, Ore.: Timber Press.

Tuckerman, E. 1865. *New-England's Rarities by John Josselyn*. Boston, Mass.: William Veazie.

U.S. Department of Agriculture. 1971. *Common Weeds of the United States*. New York: Dover.

U.S. Department of Agriculture. 2008. *Plants Database*. Available on the Internet at http://plants.usda.gov/

Uva, R. H., J. C. Neal, and J. M. DiTomaso. 1997. *Weeds of the Northeast*. Ithaca, N.Y.: Cornell University Press.

Vaughn, J. G., and C. A. Geissler. *The New Oxford Book of Food Plants*. Oxford: Oxford University Press.

Vessel, M. F., and H. H. Wong. 1987. *Natural History of Vacant Lots*. Berkeley: University of California Press.

Weed Science Society of America [WSSA]. 2007. *Composite List of Weeds*. Available on the Internet at http://www.wssa.net/Weeds/ID/WeedNames/namesearch.php/.

Weiss, J., Wolfgang Burghardt, Peter Gausmann, Rita Haag, Henning Haeupler, Michael Hamann, Bertram Leder, Annette Schulte, and Ingrid Stempelmann. 2005. Nature returns to abandoned industrial land: monitoring succession in urban-industrial woodlands in the German Ruhr. In *Wild Urban Woodlands*, ed. I. Kowarik and S. Körner, pp. 143–162. Berlin: Springer-Verlag.

Whitney, G. G. 1985. A quantitative analysis of the flora and plant communities of a representative midwestern U.S. town. *Urban Ecology* 6:143–160.

Wittig, R. 2004. The origin and development of the urban flora of Central Europe. *Urban Ecosystems* 7:323–339.

Zerbe, S., U. Maurer, S. Schmitz, and H. Sukopp. 2003. Biodiversity in Berlin and its potential for nature conservation. *Landscape and Urban Planning* 62:139–148.

Zipperer, W. C., S. M. Siginni, and R. V. Pouyat. 1997. Urban tree cover: an ecological perspective. *Urban Ecosystems* 1:229–246.

Ziska, L. H., Dennis E. Gebhard, David A. Frenz, Shaun Faulkner, Benjamin D. Singer, and James G. Straka. 2003. Cities as harbingers of climate change: common ragweed, urbanization and public health. *Journal of Allergy and Clinical Immunology* 111(2):290–295.

Ziska, L. H., J. A. Bunce, and E. W. Goins. 2004. Characterization of an urban-rural CO_2/temperature gradient and associated changes in initial plant productivity during secondary succession. *Oecologia* 139:454–458.

INDEX

Page numbers in **boldface** indicate the location of the main text description. Entries without a boldface page number are synonyms.